大学物理实验双语教程

主　编　史金辉　邢　健　张晓峻　朱　正
主　审　孙晶华

哈尔滨工程大学出版社

内容简介

本书是根据高等院校物理实验课程的基本要求,结合哈尔滨工程大学近年来针对留学生物理实验教学的经验,在历年来所用中文实验教材的基础上编写而成的。本书介绍了有关物理实验的数据处理知识,精选了力学、热学、电磁学、光学及近代物理共 20 个实验。

本书可作为理工科类来华留学生的物理实验教材,也可作为双语教材供中国学生选用。

图书在版编目(CIP)数据

大学物理实验双语教程/史金辉等主编. —哈尔滨:
哈尔滨工程大学出版社,2014.7
ISBN 978 - 7 - 5661 - 0853 - 1

Ⅰ. ①大… Ⅱ. ①史… Ⅲ. ①物理学 - 实验 - 双语教学 - 高等学校 - 教材 Ⅳ. ①O4 - 33

中国版本图书馆 CIP 数据核字(2014)第 162165 号

出版发行	哈尔滨工程大学出版社
社　　址	哈尔滨市南岗区东大直街 124 号
邮政编码	150001
发行电话	0451 - 82519328
传　　真	0451 - 82519699
经　　销	新华书店
印　　刷	哈尔滨市石桥印务有限公司
开　　本	787mm × 960mm　1/16
印　　张	19
字　　数	395 千字
版　　次	2014 年 7 月第 1 版
印　　次	2014 年 7 月第 1 次印刷
定　　价	40.00 元

http://www.hrbeupress.com
E-mail:heupress@ hrbeu.edu.cn

前　言
PREFACE

物理实验是我国高等理工科院校必开的一门基础课程，是培养学生实验能力和科学素质的基石。随着国际交流日益频繁，越来越多的留学生到中国各大高校交流学习。但是国内适合于留学生物理实验教学的双语教材并不多见。因此，我们根据教育部2008年颁发的"理工科类大学物理实验课程教学基本要求"，结合哈尔滨工程大学近年来对留学生物理实验教学的经验，在历年来所用中文实验教材的基础上，吸收国外经典物理实验的内容，编写了这本大学物理实验双语教材。

本书主要分为两部分内容：第一部分介绍了误差理论、不确定度的概念及测量结果的评定，数据处理的基本方法和常用物理实验仪器介绍；第二部分共精选了20个实验，包括基础实验、设计性实验和综合性实验，涉及力学、热学、光学、电磁学和近代物理的内容。本书可作为理工科类来华留学生的物理实验教材，也可作为双语教材供国内学生选用。

参加本书编写工作的有：史金辉(实验3、7、10、13、15、17)，邢健(实验4、8、9、11、16)，张晓峻(实验12、14、18～20)，朱正(绪论、实验1、2、5、6)。全书由孙晶华老师主审，修改定稿。

本书的编写离不开物理实验中心教师的大力支持，凝聚着全体任课教师的共同努力和辛勤劳动。在编写过程中，参考了一些兄弟院校的教材及国内的文献资料，在此谨致深切的谢意。

由于编者的水平有限，书中难免有错误和不妥之处，敬请读者批评指正。

编　者
2014年5月于哈尔滨

Introduction		1
Experiment 1	Measurement of the Moment of Inertia by Torsion Pendulum	27
Experiment 2	Measurement of Young's Modulus of Metallic Wire by Tension Method	41
Experiment 3	Measurement of Surface Tension Coefficient of Liquid	51
Experiment 4	Determination of Ratio of Specific Heat of Air	63
Experiment 5	Mapping of Electrostatic Field by Imitative Method	73
Experiment 6	Measurement of Solenoidal Magnetic Field by Hall Effect Method	84
Experiment 7	Electronic Deflection of Electron Beam and Measurement of the Specific Charge of Electron	103
Experiment 8	Determination of Focal Length of Thin Lens	115
Experiment 9	Adjustment and Application of Spectrometer	126
Experiment 10	Experiment of Equal-thickness Interference	139
Experiment 11	Measurement of Sound Velocity	149
Experiment 12	Measurement of Metal Electronic Work Function	170
Experiment 13	Michelson Interferometer	178
Experiment 14	Franck-Hertz Experiment	193
Experiment 15	Photoelectric Effect Experiment	203
Experiment 16	Fundamental Experiment of Optical Fiber Sensors	216
Experiment 17	Optical Communication Experiment	240
Experiment 18	Modification and Calibration of Electricity Meters	261
Experiment 19	Measuring Resistance by Assembly Bridge	277
Experiment 20	Assembly Telescope and Microscope	287
参考文献		297

Introduction to Physical Experiment

1.1 Basic concepts of measurement and error

Physical quantities are often needed to be measured and usually quantitative relations between physical quantities are needed to be found out in the process of production, life and scientific research. If related information of their big or small quantity values is to be obtained, people need have the aid of a tool or instrument, in this way there is a problem of confidence level about measured values that are obtained. So concepts of the measurement and error are introduced.

1. Basic concept of measurement

So-called measurement is to compare the physical quantity to be measured with a uniformity quantity as a standard to obtain multiple relation between them, and the multiplication result is the measured value to be measured. The uniformity quantity as the standard is called as "Unit". For example, use a standard meter scale to measure the track length of sports ground, and the length from starting point to finishing point is just 100 standard meter scale length, so the length of track is equal to $1 \text{ m} \times 100 = 100$ m, here 100 is as the multiple of unit 1 m. As required, for a measured result usually there are a few expression methods: Values having confirmed unit, curves on confirmed coordinate, figure drawn out as a certain proportion and collected waveforms or images etc.

2. Classification of measurement

Measurement is usually divided into two classes by methods: So-called direct measurement means the measurement process by using measurement tools or measurement instruments to directly measure or read out measured values. As shown in Fig. 1.1 - 1, use a meter scale to measure the height of a child directly, use a thermometer to measure temperature, use a stopwatch to measure time, and use a multi-meter to measure electric current, voltage and resistance directly etc. Direct measurement also includes single direct measurement and multi direct measurement. And indirect measurement means the process that measurement of some physical quantities can not be completed only by direct measurement, and the required measured values can be obtained by calculation using formulas after several physical quantities must be directly measured. As shown in Fig. 1.1 - 2, for the measurement of density ρ of a cylinder, first we can directly measure height h and diameter d of the cylinder and weigh its mass M, and then substitute them into formula to calculate density ρ of a cylinder. In physical measurement, most of them belongs to indirect measurement, but indirect measurement is basic to all physical measurements.

Fig. 1.1 - 1   Direct measurement diagram

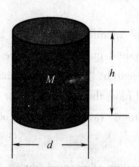

Fig. 1.1 - 2   Indirect measurement diagram

Based on different measurement condition, measurement can also be divided into equal observations and unequal observations. Equal observations means multi repeat measurement containing five basic elements such as measurement instrument, measuring personnel, measurement method, measurement environment and the measured participating measurement that have no change. Equal observations is also called as repeatability measurement.

During multi repeat measurement, it is not possible for five elements for measurement to keep unchanged absolutely, so equal observations is an ideal concept. When equal observations is made to a fixed measured object, it is allowable for the obtained measured data to have a big or small change in a range, we can not judge which value is more nearly to the measured true value, so those values are only equally dealt with. The confidence level for the measured data is same.

Unequal observations means the measurement made that five elements participating measurement except the measured object, all or any of the other four elements changes. Unequal observations is also called as reproducibility measurement.

During measuring, because of measurement conditions changed, for example, different observers measure the measured object for different times under the different conditions using different instruments and different methods, factors affecting and determining measured results are different, and confidence level for the measured data is different. Unequal observations is often used for high accuracy measurement.

3. Basic concept of error

It is not possible to realize absolute accuracy by any measurement instrument, measurement method, measurement environment and observation of measuring operators so that it is not possible to avoid error occurrence with measurement. When analyzing various errors produced possibly in measurement, their affection shall be eliminated as far as possible, and the errors in measured results that can not be eliminated are estimated, which should be the problems covered in physical and scientific experiments, so error theory and data process methods appear.

The quantity value measured is quantitative sign of a characteristic of objective things, but they can not reflect objective existence completely accurately, only limitlessly trend to reflect a quantity value characteristic of objective things. True value, abbreviated to true value, referred to as x_0, means the quantity value of a measured object, existed objectively under a certain condition. Because measurement error exists generally, so it is not possible to get the truth value of measured object by measurement. Only optimum estimated value or called as measured value of truth value, referred to as x can be obtained by measurement.

The difference of measured value and true value is called as measurement error, expressed as

$$\Delta x = x - x_0 \qquad (1.1-1)$$

In which, Δx is called as absolute error of measurement.

Because absolute error only expresses measured error big or small, it is impossible to express accurate precision level of measurement that has been done by use of absolute error when the measured object with different order of magnitude is measured. For example, different physical quantities of two order of magnitudes are separately measured, if the obtained measured absolute errors are same, but their measured precisions are not same, the one with high order of magnitude, measurement precision is high, the one with low order of magnitude, measurement precision is low. Therefore, comparison is done to measured precision or the precision of measurement instrument, relative error must be used as qualification of mutual comparison.

In order to evaluate more accurately measured results good or not, people have introduced concept of relative error, which is defined as the ratio of the absolute error to the true value, and expressed by percent, that is

$$E_r = \frac{\Delta x}{x_0} \times 100\% \qquad (1.1-2)$$

It is clear that relative error is a dimensionless value. The less the relative error of measured result is, which shows the measured result is closer to true value.

4. Source of error

During measuring, generally source of error can be summarized as the following reasons:

(1) Method error: It is a measurement error introduced because of the used measurement principle or measurement method itself. Main reason of this error source: Research of related knowledge about measured object is not enough, effects of different factors can not be considered completely, it is limited by objective conditions and technical level, the used measurement principle itself is approximate, or some factors that have real function in measurement are ignored, original state of measured object is damaged using contact measurement method, and static measurement method is used to measure a dynamic object etc.

(2) Instrument error: Means the error produced by effects of inherent various factors on the used measurement equipment and instrument themselves during measurement, for example, accuracy, sensitivity, minimum division value and stability good or not etc. This depends on factors such as structure of measurement device, design, performance of elements and device, performance of parts materials, manufacturing and technical level of fitting etc. And when designing and manufacturing various measurement devices, only a certain condition and real requirements are met. There is always a gap from ideal requirements, so a certain error always exists in measured value during measuring.

(3) Environmental error: Because surroundings affect measurement, error is produced in measurement. These effect factors always exist outside of measurement system, but have function on measurement system directly or indirectly, for example, temperature, humidity, atmospheric pressure, electrical field, magnetic field, mechanic vibration, acceleration, gravitation, sounding, illumination, dust, various rays or electromagnetic wave etc. In different measurements, the factors may have different degree of effects on measurement. They can affect the measurement error produced by measurement system, and sometimes can also cause change of the measured object, if it is serious, it can cause measurement equipment damaged or make measurement difficult continue. In order to distinguish environmental error and instrument error,

generally the so-called standard environment condition (reference condition) is determined artificially, or operation conditions of measurement instrument are specified on product placard and operation instruction. The measurement error produced in measurement made under reference condition is considered as the inherent error of measurement instrument (instrument error). If measurement instrument works in the environment specified without reference condition, the effects of environment factor make measurement error increase. This measurement error increment is called as additional error of instrument, which is environment error. Therefore when an instrument carries out measurement in meeting condition specified, the obtained measured value error, big or small, must not exceed the error value given on placard or in instruction. Some instruments also give the changed environment error value with environment condition varied.

(4) Subjective error: Subjective error is also called as personnel error. It is the error caused by quality condition of operators who carry out measurement, there is a class of error that is difficult to avoid, for example, observation error caused by resolution capability of sense organ, reaction retard, habit sense and operation level factor of measuring personnel. The other one is subjective error that must be avoided as far as possible: for example, measured error caused by reading, record, and calculation mistake caused by careless measuring operator, or operation fault.

The above-mentioned source of four measurement errors is summarized by four links participating measurement, that is personnel, equipment, method and condition. In specific measurement, the level of various factors affecting measurement is different, and even measurement error caused by a factor is small so that it is ignored.

5. Classification of error

In order to know measured error further systematically and summarily, classification method for measured error generally used is introduced now, where classification is made in accordance to characteristic and features of measured error. Measured error can be divided into three big classes: Accidental error, system error and gross error.

(1) Accidental error: During measuring, measured error of accidental nature is resulted from some affection of accidental factors, which is called as accidental error. Accidental error is also called as fortuitous error. It is difficult to calculate this error size and direction (positive and negative error). Even if under the same condition as far as possible to measure a specified physical quantity repeatedly, and after all obviously regular deviations are eliminated and corrected as far as possible, the measured value obtained every time is always fluctuation changed accidentally within a certain range.

Both statistical theory and experiment prove that in the most of physical measurements, when repeated measurement times is enough, accidental error obeys normal distribution law. If Δx is used to express accidental error of measured value for a physical quantity, $p(\Delta x)$ is probability the density function of accidental error. Its mathematical expression is

$$p(\Delta x) = \frac{1}{\sigma \sqrt{2\pi}} e^{-(\Delta x)^2/2\sigma^2} \qquad (1.1-3)$$

Where, $\sigma = \sqrt{\frac{1}{n} \sum_{i=1}^{n} (x_i - x_0)^2}$ is standard error of measurement.

For an individual, accidental error is the error produced by any measurement during repeated measurement, which is irregular, can not be controlled, and is difficult to eliminate by experiment method. For total, that is the measured values obtained by many times of measurement, and accidental error obeys a certain statistical law(Normal distribution or Gaussian distribution). Therefore for accidental error, probability statistics method can be used to deal with, i. e. characteristic quantity—standard error σ is used to express. The more the measurement times n, the less the standard error σ, that is, the less the accidental error is. It is obvious that accidental error is related with measurement times. Measurement times is increased to reduce accidental error effect on measurement result.

In formula (1.1-3), measured standard error σ is a basic index to evaluate accidental error, and its values are decided by factors such as measurement tool, instrument, measurement environment, measurement personnel and measured object etc. For the same object measured, after measurement system (including the above factors) is determined, the values of standard error σ can be determined, too. When the measurement system is not same (if other standard device and other instrument is used, or measurement environment is changed, or other measurement method is used etc.), the value σ taken is different. Therefore after measurement system is determined, its standard error is a determined constant followed with it. At the moment, in the formula (1.1-3), there is only an accidental error Δx, one variation, the curve is only determined. Obviously, when σ is different, the form of corresponding curve is different, either. In Fig. 1.1-3, distribution curves of normal distribution probability density function $p(\Delta x)$ for $\sigma = 0.5, \sigma = 1$ 与 $\sigma = 2$ are given separately in Fig. 1.1-3. It can be seen from the diagram, the less the value σ is, the steeper

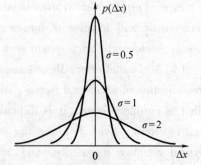

Fig. 1.1-3　Normal distribution curve of accidental error

the normal distribution curve is, and error distribution trends to concentrate.

In accordance with probability statistics theory, formula (1.1-3) is integrated to obtain

$$P = \int_{-\infty}^{+\infty} p(\Delta x) \mathrm{d}(\Delta x) = \int_{-\infty}^{+\infty} \frac{1}{\sigma \sqrt{2\pi}} e^{-\frac{(\Delta x)^2}{2\sigma^2}} \mathrm{d}(\Delta x) = 1 \qquad (1.1-4)$$

It shows that probability of accidental error of measurement set in $(-\infty, +\infty)$ interval is 1, i.e. the area under probability density distribution curve is 1. So probability density distribution function is used to calculate the probability of accidental error measured in a time set in $(-\sigma, +\sigma)$ interval, which is

$$P_\sigma = \int_{-\sigma}^{+\sigma} p(\Delta x) \mathrm{d}(\Delta x) = 0.683 \qquad (1.1-5)$$

In the same principle, probability of accidental error for measurement in a time set $(-2\sigma, +2\sigma)$ and $(-3\sigma, +3\sigma)$ interval are separately

$$P_{2\sigma} = \int_{-2\sigma}^{+2\sigma} p(\Delta x) \mathrm{d}(\Delta x) = 0.954 \qquad (1.1-6)$$

$$P_{3\sigma} = \int_{-3\sigma}^{+3\sigma} p(\Delta x) \mathrm{d}(\Delta x) = 0.997 \qquad (1.1-7)$$

When percentage is used to express, they are 68.3%, 95.4% and 99.7% separately, and their corresponding curves' areas are shown in Fig. 1.1-4.

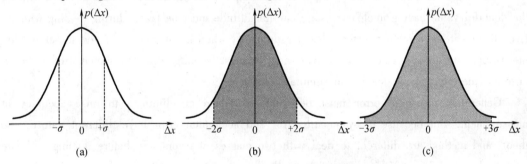

Fig. 1.1-4 Probability of error set in a interval
(a) $P_\sigma = 68.3\%$; (b) $P_{2\sigma} = 95.4\%$; (c) $P_{3\sigma} = 99.7\%$

As stated above, characteristic of accidental error can be summed up as three aspects: Accidental, produced in measurement, related with times of measurement. At equal precision, increment of measurement times can reduce accidental error effect on measured result. Accidental error obeying normal distribution has the following characteristics: Symmetry, i.e. appearance probability of positive error and negative error with equal absolute value is equal. Unimodality, i.e. probability of error appearance for small absolute value is high, and probability of error appearance for big absolute value is very low. Boundedness, probability of very big positive or negative error appearance is almost zero.

(2) Systematic error: When its size and direction have a certain rule, the measurement error is called as systematic error, also is determined error. In accordance with known change law, it can be further divided into: Fixed and unchanged systematic error is called as constant systematic error, and systematic error changed in accordance with known law is called as variable-value systematic error. Therefore systematic error can be tried to eliminate.

The determined error above is more typical systematic error according to its knowability, which is generally called as determined systematic error. For the opposite to this, if systematic error has a certain accidental nature, systematic error is called as undetermined systematic error.

(3) Gross error: Error which obviously distort measured value is called as gross error. The error is caused by wrong operation, wrong reading and wrong record etc., i. e. due to carelessness or fault, so it is also called as carelessness and fault error.

Gross error, seen from absolute value, is much greater than general systematic error value or accidental error at similar condition. Therefore the measured value with gross error has a large difference from normally measured value, so it is called as abnormal value or doubtful value.

The method to deal with gross error is to eliminate it from the measured data directly. But for the doubtful value having unclear reason, careful attitude must be taken during dealing with it. Even though it has a more effect on the measurement, when it can not be judged unbelievable, never eliminate it easily against subjective wish. It must be judged in accordance with a certain criterion, the data can be decided to eliminate finally.

Generally, measured error must be tried to reduce or eliminate to raise accuracy of measurement. But because of systematic error and accidental error have different nature, the theory and method are different to deal with two classes of errors. So before dealing with the measured error, it must be distinguished whether it is systematic error or accidental error according to error nature.

1.2 Effective figure

In actual measurement, numbers can be divided into two classes based on the digit occupied by figure that is effective or not. One effective digit is unlimited number, and this number is mostly the result of pure mathematical calculation, for example, digits $\sqrt{2}$, π, 1/3 etc., for which digits taken, as required, are all effective. The other effective digit is limited figure, and this number is mostly connected with actuality. It is not possible to determine freely its effective digit only by mathematical calculation, and the quantity value to be expressed or the accuracy possessed are expressed properly based on actuality. The effective digit for this number is limited by factors such as the precision that original data can reach, the technical level to obtain data, and the theory on which to obtain data etc. During processing the measured data, it is very important to grasp related knowledge of effective figure.

1. Basic concept of effective figure

A numerical value is composed of figures, other figures are reliable values or explicit values except that the last digit is not an explicit value or a doubtable value, and all figures forming the number including last digit is called as effective figure, i. e. accurate value plus a digit of pure doubtable number forms digit of effective figure.

As shown in Fig. 1.2 − 1(a), a steel ruler whose length is 15 cm and minimum division value is 1 mm is used to measure length of an object, and the read length is 7.26 cm. In this read data, 7.2 is read directly from the scales on the steel ruler, and the last digit 6 is estimated by a measurer, which is doubtful, because different measurers' estimated read result may be different, and it is called as pure doubtful figure. So it is can be said that a measured result has three-digit effective figures. Is it possible to try to think if the next digit of the 6 can be read out? Therefore when using a steel ruler to measure length of an object, if the measured result is expressed using mm as a unit, only the last digit behind decimal point can be read out, and then the digit of effective figure is determined.

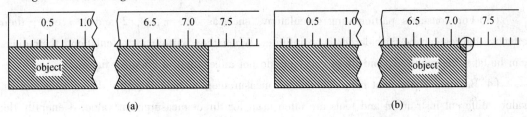

Fig. 1.2 −1 **Measuring object length by a ruler object**
(a) General; (b) Special

2. Stipulation and instruction of effective figure

(1) Digit of effective figure is not related with the position of decimal point, and unit change does not affect digit of effective figure. For example, length of an object $L = 12.01$ cm =

0.120,1 m = 0.000,120,1 km, they are all four-digit effective figure. It can be seen from the example that decimal system unit change can not affect the digit of effective figure, and at the moment, generally scientific counting process can be used, for example, the length of object above can be expressed as $L = 12.01 \times 10^{-4}$ km, and it is still four-digit effective figure because power exponent of "10" is not counted in digit of effective figure.

(2) Figure "0", has special characteristic in effect figure. First look at the three expression methods, which seems different, of the above length L of an object. Where "0" appears in different positions of data. The "0" in 12.01 is obviously counted in the digit effective figure, and the "0" before "1" in the figures 0.120,1 and 0.000,120,1 is obviously not counted in the digit of effective figure, i.e. when calculating digit of effective figure, the first "0" before nonzero figure is not counted in digit of effective figure. Let's look at the measurement as shown in Fig. 1.2-1(b), when measuring, one end of an object just aligns with a scale line on the ruler. At the moment if the measured result can be recorded as 7.2 cm or not, the answer is not allowable. It must be counted as 7.20 cm. As viewed from mathematics, 7.2 cm and 7.20 cm are the same two values. It seems that it is not necessary to keep the "0" in the latter. But as viewed from measurement error and effective figure, both of them is not the same completely, the "2" expressed as 7.2 is already pure doubtable number, i.e. the measurement precision of measurement tool used is 0.1 cm, and the "0" expressed as 7.20 cm is a pure doubtable number, i.e. the measurement precision of measurement tool used is 0.01 cm. Therefore during recording experiment data, it is careful that when the last digit behind decimal point is "0", which can not be freely given up, because it stands for the precision of measured tool.

(3) For constants participating calculation, such as $\sqrt{2}$, π, e, 2 and 1/3 etc., their effective figures can be considered to be infinite. When they participate calculation, their digits can be taken as required, and these constants do not affect digit of effective figure.

(4) Because experiment result of indirect measurement is calculated by direct measurement value, different instrument and tools are often used for direct measurement value. Generally the obtained digit of effective figure is different for each direct measurement value, and there is a problem of acceptance and rejection of effective figure during calculation. The followings are examples to show the method how to accept and reject digit of effective figure in the process of " + " " − " " × " and " ÷ " calculations.

3. Basic calculation rules of effective figure

" + " and " − " calculation process is as follows: The figure with point over its head is a pure doubtable number, the principle is no matter what number, only pure doubtable number participates calculation, and the result of calculation must be pure doubtable number, but the final result only keeps a pure doubtable number. Especially pay attention to: additive calculation may make digit of effective figure increase (carry), and subtractive calculation may make digit of effective figure decrease (borrow).

$$\begin{array}{r} 146.\dot{2} \\ +\ \ 2.36\dot{7} \\ \hline 148.56\dot{7} \end{array} \Longrightarrow 148.5\dot{6}\dot{7} \Longrightarrow 148.\dot{6} \quad \begin{array}{r} 30.6\dot{5} \\ -\ 4.93\dot{6} \\ \hline 25.71\dot{4} \end{array} \Longrightarrow 25.7\dot{1}\dot{4} \Longrightarrow 25.7\dot{1}$$

$\quad\quad\quad\quad\quad\quad\quad$ doubtable $\quad\quad\quad\quad\quad\quad\quad\quad\quad\quad\quad$ doubtable

" × " and " ÷ " calculation process is as follows: The calculation rules is the same with the above. For division calculation, students are requested to calculate according to the principle.

$$\begin{array}{r} 6.35\dot{6} \\ \times\ \ 30.\dot{5} \\ \hline 3178\dot{0} \\ 000\dot{0} \\ 1906\dot{8} \\ \hline 193858\dot{0} \end{array} \Longrightarrow 19385\dot{8}\dot{0} \Longrightarrow 19\dot{4}$$

$\quad\quad\quad\quad\quad\quad\quad\quad\quad\quad$ doubtable

For calculation of effective figure, the following laws can be obtained:

(1) The digit of pure doubtable number of " + " and " - " calculation result is same with the highest one of the digits of pure doubtable number of effective figure participating calculation, for example, in Example 1, "2" digit behind 146.2 decimal point is the highest, and therefore there is only one digit behind decimal point of calculation result, and Example 2 is same with the principle.

(2) The digit of effective figure of " × " and " ÷ " calculation result is generally same with the least one of digits of effective figure participating calculation, for example, in Example 3, four-digit effective figure is multiplied by three-digit effective figure, and the result is three-digit effective figure.

(3) The digit of effective figure of involution and evolution calculation result is same with the digit of effective figure of its bottom.

(4) For exponent, logarithm and trigonometric function etc., the digit of effective figure of calculation result can be determined by its variation. For example, an included angle measured by experiment is $19°35'$, and the last pure doubtable number is $5'$, converted into degree and expressed as $19.58°$, so the result calculated by computer is $\sin 19.58° = 0.335,122,7\cdots$. How to judge which one is pure doubtable number? Generally we use this method, i.e. a number close

11

to it is taken to calculate, for example, calculating sin 19.59° = 0.335,287,1⋯, it is found that by comparison of its calculated result, the fourth digit behind decimal point is different, so sin19.58° = 0.335,1 can be taken, in which "1" is the last digit of pure doubtable number.

As stated above, as expression of effective figure for experiment result, only one digit of pure doubtable number is in principle. But in the process of dealing with experiment data, it is determined according to actual condition for the final result that how many digits of effective figure is to keep. Related rules can be given in the section of estimation of uncertainty of measurement.

1.3 Estimation of measured result uncertainty

The purpose of measurement is to obtain the true value of the measurement, but it is difficult to determine the true value of the measurement because of existence of measurement error, for its measured result, only an approximately estimated value of a true value (i.e. so-called optimum estimated value) and an error range used to express degree of approximate. Then the concept of "uncertainty of measurement" is introduced, and the expression of uncertainty of measurement is used to quantify and evaluate measurement level or quality.

1. Mean arithmetical value is the optimum estimated value of true value

General error theory considers that for many times of measurement, mean arithmetical value is optimum estimated value of true value. Mean arithmetical value for many times of measurement is defined as: At system error eliminated, suppose a true value is x_0 physical quantity for n times of measurement, the measurement value obtained are separately $x_1, x_2, x_3, \cdots, x_n$. The error of measured value for any time can be expressed as

$$\Delta x_i = x_i - x_0 \quad (i = 1, 2, 3, \cdots, n) \quad (1.3-1)$$

For formula (1.3-1) summation, we can obtain

$$\sum_{i=1}^{n} \Delta x_i = (x_1 + x_2 + x_3 + \cdots + x_n) - nx_0 \quad (1.3-2)$$

Both sides of formula (1.3-2) are divided by n to obtain

$$\frac{1}{n}\sum_{i=1}^{n} \Delta x_i = \frac{1}{n}\sum_{i=1}^{n} x_i - x_0 \quad (1.3-3)$$

In accordance to symmetry of statistic law for accidental error, when $n \to \infty$ there is

$$\lim_{n \to \infty} \sum_{i=1}^{n} \Delta x_i = 0 \quad (1.3-4)$$

$$x_0 = \frac{\sum_{i=1}^{n} x_i}{n} = \bar{x} \qquad (1.3-5)$$

This shows that when measurement frequency is enough, mean arithmetical value is closest to true value. Actually when carrying out measurement of limited number of times, mean arithmetical value is also optimum estimated value of true value. Therefore in actual measurement, generally mean arithmetical value of measurement value is used to express the experiment result of direct measurement for many times.

2. Standard deviation of measurement value

Based on statistics theory, for the measurement of limited number of times, mean arithmetical value for measurement can be used to substitute true value, and standard deviation is used to substitute accidental error of measured value, and standard deviation of measured value is defined as

$$S = \sqrt{\frac{\sum_{i=1}^{n} (x_i - \bar{x})^2}{n-1}} \qquad (1.3-6)$$

Standard deviation S expresses an estimated value of standard error σ for measurement of any group of limited time. The physical meaning is that if accidental error of measurement for many times obeys Gaussian distribution, the possibility (probability) of measured value error set in the interval from $-S$ to $+S$ is 68.3% for any group of measurement for limited number of times.

3. Standard deviation of mean arithmetical value

Average value for measurement of limited numbers of times of a physical quantity is also a random variable. That is for this physical quantity, different groups of measurement for limited number of times is carried out, and generally mean arithmetical value for each group is also different, and there is difference among them. Therefore there is so-called standard deviation of the mean arithmetical value, and symbol $S_{\bar{x}}$ is used to express. The relation of $S_{\bar{x}}$ and S can be proved as

$$S_{\bar{x}} = \frac{S}{\sqrt{n}} = \sqrt{\frac{\sum_{i=1}^{n} (x_i - \bar{x})^2}{n(n-1)}} \qquad (1.3-7)$$

It can be seen that in the group of equal observations for measurement of n times, standard deviation of mean arithmetical value is less than that of standard deviation of single group of measurement. When the higher the number of times n of measurement, the closer to the true value measured the mean arithmetical value is, and the higher the measured precision will be. The probability meaning is still the probability of true value to be measured set in the interval $(\bar{x} - S_{\bar{x}}, \bar{x} + S_{\bar{x}})$, which is 68.3%, the probability in the interval $(\bar{x} - S_{\bar{x}}, \bar{x} + S_{\bar{x}})$ is 95.4%, and the probability in the interval $(\bar{x} - 3S_{\bar{x}}, \bar{x} + 3S_{\bar{x}})$ is 99.7%.

In the process of actual measurement, generally measurement times is limited, so measurement data and measurement distribution deviate from normal distribution in theory, as shown in Fig. 1.3 – 1, i. e. so-called t distribution. The solid line in the figure expresses normal distribution in theory, and obviously the curve of t distribution deviates from theoretical curve. Distribution coefficient $t_p(n)$ of t distribution function is a quantitative value related with measurement times t and fiducial probability P, and their values are shown in Table 1.3 – 1. Therefore the standard deviation of mean arithmetical value given in formula (1.3 – 7) is needed to correct in actual application, i. e. $S_{\bar{x}}$ is multiplied by $t_p(n)$.

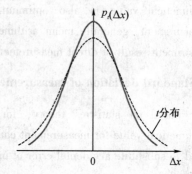

Fig. 1.3 – 1 t distribution

Table 1.3 – 1 Distribution coefficient of different fiducial probability P, number of times of measurement n

$t_p(n)$ P \ n	3	4	5	6	7	8	9	10
0.68	1.32	1.20	1.14	1.11	1.09	1.08	1.07	1.06
0.95	3.18	2.78	2.57	2.45	2.36	2.31	2.26	2.23
0.99	5.84	4.60	4.03	3.71	3.50	3.36	3.25	3.17

4. Definition of measuring uncertainty

Definition of measuring uncertainty is to characterize dispersivity of measured value, and is a parameter connected with measured result.

The word of uncertainty means doubtable degree. In a broad sense, uncertainty of measurement shows doubtable degree of measured result correctness. So uncertainty of measurement also has different-formed definition: "Measure of possible error for the measured estimated value given by measured result" "Evaluation of range in which characterizes measured true value is" "Because of existence of measurement error, the degree of uncertainty for measured result". All the definitions of uncertainty of measurement do not have an essential distinction, and their evaluation methods are same and expression forms are same, too.

A complete measured result must include two parts, optimum estimated value of the measured value and uncertainty of measurement. For example, the measurement result of measured x_0 is $x \pm \Delta$, its measured value x is optimum estimated value of x_0, and Δ is uncertainty of measurement of measured value x. Measured result can also be developed to express as ($x + \Delta$, $x - \Delta$). Obviously the one expressed by measured result of the measured x is not a confirmed value, and it characterizes the evaluation of range in which the measured true value is.

5. Classification of uncertainty

Evaluation of measured result of experiment usually is related with factors, and these factors form different components of uncertainty. In physical experiment in the university, generally the estimation for uncertainty of measurement only considers two classes of components, and then they are combined as the final uncertainty of experiment result. The two classes of uncertainty are separately:

Class A uncertainty—means in repeated equal observations of many times, uncertainty components $\Delta A_i (i = 1, 2, \cdots)$ estimated by using statistical method can be used.

Class B uncertainty—The components $\Delta B_i (i = 1, 2, \cdots)$ estimated by using other non-statistical method, some system errors that are not possible to eliminate, such as instrument error is one of typical Class B components.

The final uncertainty Δ is a combination of each component of the two classes of uncertainties. Δ_{A_i} and Δ_{B_j} are independent to each other, and have same fiducial probability. And uncertainty Δ is appointed to combine to express

$$\Delta = \sqrt{\sum \Delta_{A_i}^2 + \sum \Delta_{B_j}^2} \qquad (1.3-8)$$

In actual application, generally in Class B uncertainty in formula (1.3-8), only a component of instrument error is included, i.e. other components are considered as zero, and in Class A uncertainty, only contribution of $S_{\bar{x}}$ is considered, so formula (1.3-8) can be rewritten as

$$\Delta = \sqrt{[t_p(n) \cdot S_{\bar{x}}]^2 + \Delta_{仪}^2} \qquad (1.3-9)$$

Based on the principle of uncertainty combination, each component must have same fiducial probability P. It can be seen from Table 1.3-1 that if $t_p(n) = \sqrt{n}$ is taken, when number of times of measurement $n > 5$, its fiducial probability is greater than 95%. $\sqrt{n}$ is used as correction coefficient, $t_p(n) \cdot S_{\bar{x}} = S$ is found, therefore in actual data process, and for the purpose of simple and convenient operation, generally calculation of S is used to substitute $S_{\bar{x}}$, so combined uncertainty is rewritten as

$$\Delta = \sqrt{S^2 + \Delta_{仪}^2} \qquad (1.3-10)$$

6. Uncertainty estimation of direct measured result

In actual measurement, only when number of times is $n > 5$, correction is not considered, and formula (1.3-10) is used to estimate uncertainty of measured result. Then measured result is finally expressed as

$$x = \bar{x} \pm \Delta \,(\text{unit}) \quad \text{or} \quad x = \bar{x}(1 \pm \frac{\Delta}{\bar{x}} \times 100\%)\,(\text{unit})$$

$$E_r = \frac{\Delta}{\bar{x}} \times 100\%$$

7. Expression of measured result

During data operation, data must be calculated in accordance with the operational rule for effective figure, and the last digit is processed to the round-off principle. In order to avoid early round-out, complementary error is introduced. During operation, effective figure must keep one more digit, but never keep more progressively. For calculation and result of uncertainty, 2 digits of effective figure is always taken, and the effective digit of measured result must be aligned with the last digit of uncertainty. For relative error, two digits of effective figures are taken and expressed using percentage.

For example: A spiral micrometer is used to measure diameter D of a wire, where the measured data is

$$D/\text{mm}: 1.516, 1.519, 1.514, 1.522, 1.513, 1.523, 1.517$$

so the calculated result is $\bar{D} = 1.517,7$ mm (one more digit is kept), $S = 0.003,8$ mm, $S_{\bar{D}} = 0.001,4$ mm, Class A uncertainty of calculation is

$$\Delta_A = t_P(n) \cdot S_{\bar{D}} = \sqrt{7} \times 0.001\,4 = 2.65 \times 0.001\,4 = 0.003\,7 \text{ mm}$$

It clear that S is a very good approximation for Class A uncertainty for limited number of times of measurement.

For Class B uncertainty, here only instrument error component $\Delta_{仪}(\Delta_I)$ is considered, when a spiral micrometer is used to measure, first corrected value of zero point is read, and then measured value, in this way twice of readings for the spiral micrometer has been taken, so in accordance to the estimation rule of uncertainty, Δ_B calculation method is

$$\Delta_B = \sqrt{\Delta_{仪}^2 + \Delta_{仪}^2} = \sqrt{0.005^2 + 0.005^2} = 0.0071 \text{ mm}$$

So combined uncertainty is

$$\Delta = \sqrt{S^2 + \Delta_B^2} = \sqrt{0.0038^2 + 0.0071^2} = 0.0081 \text{ mm}$$

Finally measured result is expressed as

$$D = \overline{D} \pm \Delta = (1.5177 \pm 0.0081) \text{ mm}$$

$$E_r = \frac{\Delta}{\overline{D}} \times 100\% = \frac{0.0081}{1.5177} \times 100\% = 0.53\%$$

Note: The complete expression of experimental result includes average value of measured values, uncertainty (principle of last digit alignment), unit of measured value and relative error.

For once direct measurement, generally $\Delta_{仪}(\Delta_I)$ is taken as the uncertainty of measured result, and for $\Delta_{仪}(\Delta_I)$, generally one effective digit is taken and expressed as

(Measured result) = (Measured value) $\pm \Delta_{仪}$

The measurement instrument $\Delta_{仪}(\Delta_I)$ often used in experiment is shown in Table 1.3-2. For instrument error of digital instrument, the minimum indicated value of last digit is taken, and for measurement tool with scale mark such as meter scale, thermometer and pointer instrument, 1/2 of minimum scale is taken to a general rule, and for reading instrument with vernier, minimum division value is taken.

Table 1.3-2 Simplification and appointment of instrument $\Delta_{仪}(\Delta_I)$ often used in this course

Steel ruler, steel tape	Vernier calipers			Spiral micrometer
	1/10 mm Grad.	1/20 mm Grad.	1/50 mm Grad.	
0.5 mm	0.1 mm	0.05 mm	0.02 mm	0.004 mm
Spectrometer (1′Grad.)	Reading microscope		Michelson interferometer	Micrometer eyepiece
1′	0.005 mm		0.00005 mm	0.05 mm

8. Uncertainty estimation of indirect measurement result

In actual measurement estimation, most physical quantity is obtained by indirect measurement. When calculating indirect measurement result, each quantity value of direct measurement is substituted into principle formula to obtain measured result.

Physical quantity N given is the function for quantity of many direct measurements, i.e.

$$N = f(A, B, C \cdots)$$

In order to study question simply, we suppose to decide whether each direct measurement A, B, $C \cdots$ of indirect measurement values are independent to each other, and they have uncertainty ΔA, ΔB, $\Delta C \cdots$ of its combination separately. It can be proved by error theory that transfer formula for combined uncertainty of indirect measurement can be expressed as

$$\Delta N = \sqrt{(\frac{\partial N}{\partial A})^2 \cdot (\Delta A)^2 + (\frac{\partial N}{\partial B})^2 \cdot (\Delta B)^2 + (\frac{\partial N}{\partial C})^2 \cdot (\Delta C)^2 + \cdots} \quad (1.3-11)$$

Transfer formula for relative error is

$$E_r = \frac{\Delta N}{N} = \sqrt{(\frac{\partial \ln N}{\partial A})^2 \cdot (\Delta A)^2 + (\frac{\partial \ln N}{\partial B})^2 \cdot (\Delta B)^2 + (\frac{\partial \ln N}{\partial C})^2 \cdot (\Delta C)^2 + \cdots}$$

$$(1.3-12)$$

It is noted that for the relation of indirect measurement quantity, when sum and difference operation are main, it is more simple to use formula (1.3 – 11) for calculation, so first uncertainty is usually to obtain, then to obtain relative error. For function relation of indirect measured quantity, when multiply and division operations can be used, it is to use formula (1.3 – 12), first relative error is calculated, then for uncertainty. When measured values of A, B, $C \cdots$ etc. are substituted into calculated result, if it is single measurement, substituted into measured value, if multiple measurement, substituted into mean arithmetical value measured. For related ΔA, ΔB, $\Delta C \cdots$, single measurement is substituted into instrument error, multiple measurement is substituted into combined uncertainty.

Refer to Table 1.3 – 3 for the calculated results of some common function relations by using transfer formula.

Table 1.3 – 3 Uncertainty transfer formulas of common functions

Function forms	Uncertainty transfer formulas
$N = A + B + C$	$\Delta N = \sqrt{(\Delta A)^2 + (\Delta B)^2 + (\Delta C)^2}$
$N = A \times B \times C$	$\frac{\Delta N}{N} = \sqrt{(\frac{\Delta A}{A})^2 + (\frac{\Delta B}{B})^2 + (\frac{\Delta C}{C})^2}$
$N = \frac{A}{B}$	$\frac{\Delta N}{N} = \sqrt{(\frac{\Delta A}{A})^2 + (\frac{\Delta B}{B})^2}$
$N = aA^n$	$\frac{\Delta N}{N} = n \frac{\Delta A}{A}$
$N = \sqrt[n]{A}$	$\frac{\Delta N}{N} = \frac{1}{n} \frac{\Delta A}{A}$
$N = \sin A$	$\Delta N = |\cos A| \cdot (\Delta A)$

Example 1. Outside diameter D_2, inside diameter D_1 and the height H of metallic hollow cylinder measured using vernier calipers are listed in Table 1.3 – 4

Table 1.3 – 4 $\Delta_{\text{仪}} = 0.005$ cm (Vernier calipers of 20 gratings)

	D_2/cm	D_1/cm	H/cm
1	3.275	2.710	3.025
2	3.275	2.710	3.030
3	3.280	2.705	3.020
4	3.280	2.705	3.015
5	3.270	2.705	3.025
6	3.275	2.705	3.025

Solution: First mean arithmetical value and standard deviation of every measured value is calculated, see Table 1.3 – 5 for calculated results.

Table 1.3 – 5   List of calculated results

	D_2/cm	D_1/cm	H/cm
Averages	3.275 8	2.706 7	3.023 3
S	0.003 8	0.002 6	0.005 2

In accordance with estimation method of uncertainty, there are

$$\Delta D_2 = \sqrt{S^2 + \Delta^2} = \sqrt{0.003\ 8^2 + 0.005^2} \approx 0.006\ 3 \text{ cm}$$

$$\Delta D_1 = \sqrt{S^2 + \Delta^2} = \sqrt{0.002\ 6^2 + 0.005^2} \approx 0.005\ 6 \text{ cm}$$

$$\Delta H = \sqrt{S^2 + \Delta^2} = \sqrt{0.005\ 2^2 + 0.005^2} \approx 0.007\ 2 \text{ cm}$$

Volume of hollow cylinder is

$$V = \frac{\pi}{4}(D_2^2 - D_1^2)H = \frac{\pi}{4}(3.275\ 8^2 - 2.706\ 7^2) \times 3.023\ 3 = 8.084\ 3 \text{ cm}^2$$

In order to use transfer formula of uncertainty, first obtain logarithm of volume and partial derivative

$$\ln V = \ln \frac{\pi}{4} + \ln(D_2^2 + D_1^2) + \ln H$$

$$\frac{\partial \ln V}{\partial D_2} = \frac{2D_2}{D_2^2 - D_1^2}; \quad \frac{\partial \ln V}{\partial D_1} = -\frac{2D_1}{D_2^2 - D_1^2}; \quad \frac{\partial \ln V}{\partial H} = \frac{1}{H}$$

Substituted into transfer formula

$$E_r = \frac{\Delta V}{V} = \sqrt{\left(\frac{2D_2}{D_2^2 - D_1^2}\right)^2 (\Delta D_2)^2 + \left(\frac{2D_1}{D_2^2 - D_1^2}\right)^2 (\Delta D_1)^2 + \left(\frac{1}{H}\right)^2 (\Delta H)^2}$$

$$= \sqrt{\left(\frac{2 \times 3.275\,8 \times 0.006\,3}{3.275\,8^2 - 2.706\,7^2}\right)^2 + \left(\frac{2 \times 3.276 \times 0.005\,6}{3.275\,8^2 - 2.706\,7^2}\right)^2 + \left(\frac{0.007\,2}{3.023\,3}\right)^2}$$

$$\approx 0.015 = 1.5\%$$

$$\Delta V = V \times \frac{\Delta V}{V} = 8.084\,3 \times 0.015 \approx 0.12 \text{ cm}^3$$

Calculated result is

$$\begin{cases} V = (8.08 \pm 0.12) \text{ cm}^2 \\ E_r = 1.5\% \end{cases}$$

1.4 Processing method of experimental data

For physical experiment except measurement of physical quantities, relationship between physical quantities and variation law are often needed to study. The followings are processing methods of experimental data which are often use.

1. Tabulation method

Tabulation is a effective method to record experimental data in sequence, is an original method to use experimental data to show function relationship. Tabulation method may shows corresponding relation between related physical quantities simply and clearly and with compact form to check result for rationality at any time, find out problems, reduce and avoid mistakes, to be helpful to find out lawful connection between physical quantities for empiric formula etc. In Table 1.4 − 1 a group of measured values of resistance R and temperature t for a piece of resistance wire is listed. It is known that by value analysis, resistance value of a resistor can increase with temperature raised

Table 1.4 − 1 Relation of $R - t$

$t/°C$	20.0	25.0	30.0	35.0	40.0	45.0	50.0	55.0
R/Ω	42.8	45.7	48.8	52.0	55.1	58.3	61.3	64.1

Requirements of tabulation:

(1) Name tabulated must be given, and tabulation must be simple and clear for seeing the relation between related quantities and for processing data.

(2) In the tabulation, the meaning of symbol standing for physical quantity must be marked, and unit is given. Unit and the order of magnitude of quantity value are given in the title column of the symbol, not necessary to write to each value repeatedly.

(3) Form of tabulation is not limited, and items should be listed in accordance to special conditions. Some data specific or not much related with other items can not be listed in the table. In the table, except original data, some intermediate results and final results during calculation also can be listed in the table.

(4) If measured datum has function relation, they must be arranged in the sequence from small to big or from big to small.

(5) The data listed in the table must reflect correctly the effective figure of measured results.

2. Graphic interpretation

In most cases, it is difficult that the relations between physical quantities are expressed with a simple analytical function formula, or it is not necessary to get function relation formula, such as air temperature change chart in one year, and transistor characteristic curve etc. At the moment graphic interpretation can be used to express directly and vividly. In order to get nice and standard chart line, the following is a brief introduction of plotting procedure and cautions of a general experimental chart line, i.e. so-called three elements of plotting: Coordinate axis name (physical quantity represented), division value and unit of division value.

(1) Selection of suitable coordinate paper

Coordinate papers often include rectangular coordinate paper, logarithmic coordinate paper and polar coordinate paper etc. In physical experiment, rectangular coordinate papers with millimeter graduation are mostly often used.

Coordinate papers' size and coordinate axis's proportion must be decided in accordance with effective figure of measured data and precision requirements of result. In principle, the minimum limitation for selecting coordinate papers' size is not to lose effective figure of experiment data and contain all the experiment points, and certainly proper enlargement is possible, too.

(2) Selection of coordinate axis and its coordinate

Generally abscissa axis expresses independent variable, and vertical axis expresses dependent variable. Each axis representing physical quantity and unit is marked separately, and its division value is marked on coordinate axis in every specified and equal interval. Generally speaking, division values must be arranged along coordinate axis in the sequence from small to big or from big to small. When taking division of coordinate axis we must take care to try to make chart "fill" up the complete drawing more symmetrically as far as possible.

(3) Marking coordinate points of measured data

Symbols such as "×" "+" "⊙" and "△" etc. are used to mark data points on coordinate paper. After plotting finished, keep the symbols to distinguish.

(4) Connection of experimental chart line

Tools such as a ruler or curved drawing instrument and sharpened hard pencil etc. are used to pass through or close all the experimental points as far as possible, and smooth curve or straight line are drawn out (except correction curve of wattmeter, generally folding line is not connected to experiment points). It is not required strongly to let chart line pass through all the experimental points, but required that experimental points on both sides of chart line are distributed uniformly, and close to chart line as far as possible.

3. Descriptive geometry solution

Descriptive geometry solution is one of the important methods to process experimental data. Chart line can directly show relation between experimental datum, and mathematic relationship between two quantities can be often found out by chart line. Main points to plot are similar to graphic interpretation, but the purpose is usually to obtain quantity value of a physical quantity by plotting and shows complex functional relation using simple chart line to get its physic law etc. Simple and often used methods include straight line descriptive geometry solution and curve changing to straight line. The two methods are introduced below.

4. Steps of line descriptive geometry solution

(1) Point selection

At both ends of straight line, taking any two points $A(x_1, y_1)$, and $B(x_2, y_2)$ as shown in Fig. 1.4 – 1, it is better that its coordinate values are integral value and they are expressed by the symbols different from data points, and coordinate readings are marked near them.

(2) slope K

Linear equation is $y = kx + b$, coordinate values for two points A and B are known, so slope k can be expressed as

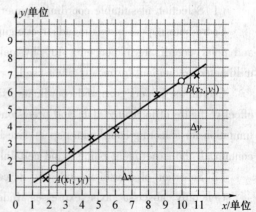

Fig. 1.4 – 1 Diagram of straight line descriptive geometry solution

$$k = \frac{y_2 - y_1}{x_2 - x_1} = \frac{\Delta y}{\Delta x} \tag{1.4-1}$$

In order to reduce relative error, the two points taken must be a little apart as far as possible within experimental data, but it is not allowable to take original experimental data.

(3) intercept b

If starting points of horizontal and vertical coordinates are zero, intercept of straight line can be read from diagram. The size is just the crossing point value of extension line of straight line and y axis, while if starting points are not zero, intercept b can be calculated using formula, i.e.

$$b = \frac{x_2 y_1 - x_1 y_2}{x_2 - x_1} \tag{1.4-2}$$

5. Method of successive difference

For function formula of linear relation, $y = ax + b$, if the change of independent variable x is equal interval, and its error is far less than the error of dependent variable y, during processing data approximate method ignoring x error can be selected to simplify problem some.

A specific method is that the datum of even group obtained by measurement is divided into two groups, front half and rear half, and then corresponding items are subtracted, and averaging is done again. As below, that experiment of Young' modulus of metallic wire measured by tension method is taken as example to show application of method of successive difference.

It is known that between the rod reading n seen in telescope and the applied force F borne on steel wire meets linear relation $F = kn$, where k is a proportionality constant. During experiment each weight mass is 0.500 kg, and experiment datum is shown in Table 1.4-2.

Table 1.4-2 Application example of method of successive difference

Measure times i	1	2	3	4	5	6	7	8	9	10
Weight mass/kg	0.500	1.000	1.500	2.000	2.500	3.000	3.500	4.000	4.500	5.000
Rod reading n_i/cm	15.95	16.55	17.18	17.80	18.40	19.02	19.63	20.22	20.84	21.47

First look at the subtracted results of successive terms, that is subtraction of the front reading from the rear reading of the reading of every 0.500 kg weight increased, then elongation quantity of steel wire of every 0.500 kg weight increased is calculated, and average is taken again, i.e.

$$\overline{\Delta n} = \frac{(n_2 - n_1) + (n_3 - n_2) + \cdots + (n_{10} - n_1)}{9} = \frac{21.74 - 15.95}{9} = 0.613\ 3\ \text{cm}$$

so proportional coefficient is

$$k = \frac{\overline{\Delta n}}{\Delta F} = \frac{0.6133}{0.500} = 1.227 \text{ cm/kg}$$

It is worth noticing that in the above calculation, not all the data are used, only two measured values of beginning and end, and this calculation method is equivalent to 10 pieces of 0.500 kg weight added once. Correct method must be that the measured data is divided into two groups, i.e. $(n_1, n_2, n_3, n_4, n_5)$ and $(n_6, n_7, n_8, n_9, n_{10})$, and then corresponding terms are subtracted for average, i.e.

$$\overline{\Delta n} = \frac{(n_6 - n_1) + (n_7 - n_2) + (n_8 - n_3) + (n_9 - n_4) + (n_{10} - n_5)}{5}$$

$$= \frac{3.07 + 3.08 + 3.04 + 3.04 + 3.07}{5} = 3.060 \text{ cm}$$

so proportional coefficient is

$$k = \frac{\overline{\Delta n}}{\Delta F} = \frac{0.6133}{0.500} = 1.227 \text{ cm/kg}$$

So all the measured data are used, which is equivalent to the results of repeated result of measurement of five times. The method of data process is specially effective to some experiments, such as Michelson interferometer, and sonic measurement in air etc. It must be noticed that if method of successive difference is used to process data, even time must be taken for its measured times. It can be seen from the above example that if measured time is odd time, inevitably there is a measured value unable to use.

6. Linear regression method

(1) Solution of optimum value of linear equation parameter (least square method)

Suppose known function as primary function form

$$y = ax + b \qquad (1.4-3)$$

Equation regression question is actually summed up that the data measured by experiment is used to determine the coefficients a and b of formula (1.4-3). It is equivalent to the solution of slope and intercept of straight line in plotting method. Because there is only one independent variable, so it is called as simple regression, also called as linear fitting. In order to determine the size of a and b, n pair of data can be

$$x_i (i = 1, 2, \cdots, n) \quad y_i (i = 1, 2, \cdots, n)$$

substituted into related formula to calculate directly optimum values of a and b.

In order to simplify the question, it is supposed that every measured value is equal accuracy, and in the measured values of x and y, only y has obviously measured accidental error. If both x and y have accidental error, it is possible to handle only a variable with small error as x.

Because of measuring in experiment, it is not avoidable that there are various errors, in general case any group of n group data can not be completely suitable to the equation (1.4-3). In order to obtain optimum empirical formula (1.4-3) parameter, n group data is substituted into equation (1.4-3), and n equation is listed below

$$v_i = y_i - (ax_i + b) \quad (i=1,2,\cdots,n) \tag{1.4-4}$$

In accordance with least square method principle, the most believable value of unknown parameters a and b can make sum of squares of each v_i as minimum value, i. e.

$$\sum_{i=1}^{n} v_i^2 = \sum_{i=1}^{n} [y_i - (ax_i + b)]^2 = (\min) \tag{1.4-5}$$

The condition making the above formula be minimum is: Their partial derivatives to a and b are all separately equal to zero

$$\frac{\partial}{\partial a} \sum_{i=1}^{n} v_i^2 = -2 \sum_{i=1}^{n} [y_i - (ax_i + b)] x_i = 0 \tag{1.4-6}$$

$$\frac{\partial}{\partial b} \sum_{i=1}^{n} v_i^2 = -2 \sum_{i=1}^{n} [y_i - (ax_i + b)] = 0 \tag{1.4-7}$$

Through arrangement of formula (1.4-7) to obtain

$$b = \frac{1}{n} \left(\sum_{i=1}^{n} y_i - a \sum_{i=1}^{n} x_i \right) \tag{1.4-8}$$

Formula (1.4-8) is substituted into formula (1.4-6), after arrangement there is

$$a = \frac{\sum x_i y_i - \frac{1}{n} \sum x_i \sum y_i}{\sum x_i^2 - \frac{1}{n} \left(\sum x_i \right)^2} \tag{1.4-9}$$

Suppose $\bar{x} = \frac{1}{n} \sum x_i$, $\bar{y} = \frac{1}{n} \sum y_i$, then

$$L_{xx} = \sum x_i^2 - \frac{1}{n} \left(\sum x_i \right)^2 = n(\overline{x^2} - \bar{x}^2)$$

$$L_{yy} = \sum y_i^2 - \frac{1}{n} \left(\sum y_i \right)^2 = n(\overline{y^2} - \bar{y}^2)$$

$$L_{xy} = \sum x_i y_i - \frac{1}{n} \sum x_i \sum y_i = n(\overline{xy} - \bar{x} \cdot \bar{y})$$

So formulas (1.4-8) and (1.4-9) can be written as

$$a = \frac{L_{xy}}{L_{xx}} \qquad (1.4-10)$$

$$b = \bar{y} - a\bar{x} \qquad (1.4-11)$$

The obtained a and b are substituted into linear equation, i.e. optimum empirical formula is obtained. Linear fitting method is widely used in scientific experiments, especially after microcomputer and calculator, calculation is reduced greatly, and calculation precision is also guaranteed, so the method is both practical and convenient. Parameters a and b calculated by this method are optimum, but it does not mean that there is no error. Because their error estimation is more complex, here is no requirement.

(2) Result inspection—correlation coefficient

When using regression method to process data, it is most difficult to select functional form. The selection of functional form is mainly based on theoretical analysis, when the theory is not clear, only the variation trend of experiment data is used to infer. In this way, for the function relation between observation quantities x and y of the same group, the function form "predicted" by different people is different. Therefore if the relation is reliable or not, inspection must be done. A general inspection method is to calculate correlation coefficient r.

In accordance to statistical theory, correlation coefficient is defined as

$$r = \frac{\sum (x_i - \bar{x})(y_i - \bar{y})}{\sqrt{\sum (x_i - \bar{x})^2 (y_i - \bar{y})^2}} = \frac{L_{xy}}{\sqrt{L_{xx} L_{yy}}} \qquad (1.4-12)$$

It can be proved that r value is always between 0 and 1. The closer the value approaches 1, showing that data points are distributed tightly arround the obtained straight line, and correlation is good. On the contrary, the closer the value approaches 0, data points disperse, and correlation is bad, showing that it is not easy to use linear regression method for the function relation, and other functions must be used to try again.

Measurement of the Moment of Inertia by Torsion Pendulum

1. Background and application

The moment of inertia is the measurement of inertia of rigid body in rotation, and the important engineering technical parameter to study, design and control the motion law of rotary object. The measurement of the moment of inertia is a part of the measurement of mass natural parameters and an important parameter in analysis of system performance of equipment. In many important industry fields, such as in space flight industry, artificial satellite, long-range rocket and tactical missile, etc., the measurement of the moment of inertia is required to confirm whether the products conform to the requirements of design and how to correct the products; in aviation industry, the measurement of the moment of inertia of aircraft is required to know the maneuverability performance of aircraft; in national defense industry, the measurement of the moment of inertia of antitank missile, rocket projectile, different cartridges, etc. is required to confirm the effects of physical parameters on initial disturbance of bullets and trajectory, etc.; and in automotive industry, the measurement of the moment of inertia and eccentricity must be made on various vehicles and rotational parts to improve their performance and extend their lives through correction of eccentricity. Therefore, it is significant to calculate the moment of inertia of calibration system by using a proper method.

The moment of inertia of rigid body is equal to the square product sum of mass points in rigid body and perpendicular distance of mass point to shaft, so the moment of inertia of rigid body is related to mass distribution of rigid body, shape and rotating shaft position. If a rigid body has a regular geometry and uniform density distribution, its moment of inertia around a specific shaft can be calculated directly, and some common moments of inertia for regular rigid bodies are shown in Fig. 1 – 1. A lot of rigid bodies with complex shapes and non-uniform density distribution are often met in engineering practice; theoretic calculation of the moment of inertia is very complex; even with the help of some calculation and analysis software, it is difficult to obtain the required precision in general; usually the experiment method is used to measure the moments of inertia. There are many methods to measure the moment of inertia, such as falling body, double-line pendulum, physical pendulum and torsion pendulum (Triple-line pendulum, metallic rod torsion pendulum, single suspension torsion pendulum, double suspension torsion pendulum and worm spring torsion pendulum), etc. For this experiment, torsion pendulum is used to measure the moment of inertia of object, worm spring torsion pendulum is used to make a body do twist swing, and the moment of inertia of object is calculated by measuring rotational cycle and other parameters.

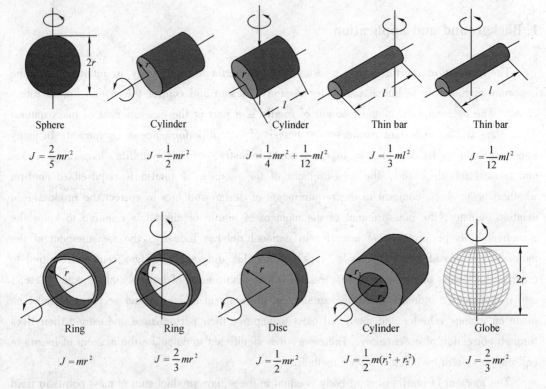

Fig. 1 – 1 Moments of inertia for some regular rigid bodies

2. Experiment instruments

The experiment equipment is shown in Fig. 1 – 2, including torsion pendulum, cycle measurement equipment (CME) for the moment of inertia, hollow metallic cylinder, plastic cylinder, solid globe, metallic long and thin bar, slide block, fixture and sliding calipers.

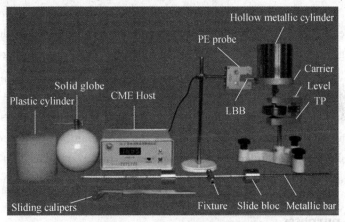

Fig. 1 – 2 The experiment equipment

Cycle measurement equipment (CME) for the moment of inertia is composed of host and photoelectric (PE) probe. The photoelectric probe is used to test light block bar (LBB) for light blocking, and to automatically judge if the set cycle number is obtained on the basis of light blocking frequency. Cycle number can be set by preset number switch. When "RESET" button is pressed, display value is "0000" seconds. When the light block bar passes through the gap of photoelectric probe for the first time, timing begins. When coming up to the preset cycle number, counting stops automatically and a four-digit number is shown. For example: "1874", the precision of measuring time is 0.01s, and the last two digits represents the value after the dot. Its unit is second, and the displayed value is 18.74 s.

The photoelectric probe comprises infrared transmitting tube and infrared ray receiving tube. Our eyes can not directly observe if the equipment works normally, but a paper card can be used to block the gap position of the photoelectric probe to check whether timer starts timing and counting stops when the preset cycle number is obtained, and if the display is "0000" when pressing the "reset" button. The photoelectric probe can not be set in strong light to prevent the photoelectric probe from being affected by strong light ray.

When testing, precautions are provided below:

(1) The base should be always in horizontal condition.

(2) The photoelectric probe is placed to the balance position of the light block bar, the light block bar must be capable of passing through two small holes in the gap of photoelectric probe, and both of them can not touch with each other.

(3) The spring twist constant K value is not a fixed constant, which is related slightly to the swing angle, and it is basically the same between 40°~90° swing angles. In order to reduce the system error caused by over high swing angle variation in test, swing angle should be basically the same, not too high or too low, and ±60° is enough when measuring swing cycle of various objects.

(4) A shaft must be entered into carrier disc, and tighten up the screws to make it fixed integral with the spring. If there is a sound when swinging or the swing angle is low obviously or stops after swinging several times, the reason is that the screws are not tightened.

(5) When solid plastic cylinder and hollow metallic cylinder are placed on the carrier disc, they must be correctly laid and not be tilt.

3. Experiment purposes

Know the construction and operation method of torsion pendulum; grasp correct operation principles of digital timer; measure the moment of inertia of object; verify parallel-axis theorem of the moment of inertia; understand some methods of data processing, and learn how to design the plan of measuring the moment of inertia of irregular objects through the test.

4. Experiment principles

Torsion pendulum equipment is shown in Fig. 1 – 3. Various objects to be tested can be installed to Shaft 1, sliced helical spring 2 is vertically mounted to Shaft 1 to produce restoring moment, 3 is level gauge to indicate the system for level, and 4 is level adjusting knob to adjust system balance.

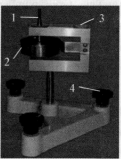

Fig. 1 – 3 **Torsion pendulum**
1—shaft; 2—spring; 3—level gauge;
4—level adjusting knob

Measurement of the Moment of Inertia by Torsion Pendulum — Experiment 1

The relationship among moment of inertia, twist constant and cycle

When the object to be tested on Shaft 1 turns a certain angle θ, the object starts twist movement back and forth around Shaft 1 with the action of spring restoring moment M. Based on Hook' law, the restoring moment M that the spring twists and produces is proportional to the turned angle θ, i. e.

$$M = -K\theta \tag{1-1}$$

Where K is twist constant of the spring.

Based on rotation law, there is

$$M = J\beta \tag{1-2}$$

Where J is the moment of inertia, β is angular acceleration.

From equation (1-2)

$$\beta = \frac{M}{J} \tag{1-3}$$

Given: $\omega^2 = \frac{K}{J}$, ignore the action of other moment of force, from equations (1-1) and (1-3)

$$\beta = \frac{d^2\theta}{dt^2} = -\frac{K}{J}\theta = -\omega^2\theta \tag{1-4}$$

The above equation shows that torsion pendulum movement has the characteristic of angular simple harmonic vibration, angular acceleration is proportional to angular displacement, and direction is contrary. The solution of equation is

$$\theta = A\cos(\omega t + \varphi) \tag{1-5}$$

Where A is angular amplitude of simple harmonic vibration, φ is initial phase angle, and ω is angular velocity.

$$T = \frac{2\pi}{\omega} = 2\pi\sqrt{\frac{J}{K}} \tag{1-6}$$

From equation (1-6), as long as either one quantity of J and K is known, another physical quantity can be calculated when the object swing cycle T is measured through experiment.

The experiment uses an object with a regular geometry, whose moment of inertia can be directly calculated by using theoretical formula based on its mass and physical dimension. Thus K value of torsion pendulum spring can be calculated. To measure the moment of inertia of other objects, only place the object to be tested to the various fixtures at the top of torsion pendulum to measure its swing cycle, and the moment of inertia of the object running around shaft can be calculated by equation (1-6).

The measurement of spring twist constant K

The measurement of spring twist constant K is shown in Fig. 1-4. Suppose the moment of inertia of metallic carrier disc around shaft is J_0', and its rotation cycle is T_0, so there is

$$T_0^2 = \frac{4\pi^2}{K} J_0' \tag{1-7}$$

The theoretic value of the moment of inertia of plastic cylinder to be tested to its centroidal axis is J_1, and the complex rotation cycle of it and carrier disc is T_1, so

$$T_1^2 = \frac{4\pi^2}{K}(J_0' + J_1) \tag{1-8}$$

Where $J_1 = \frac{1}{8} m_1 D_1^2$, D_1 is the diameter of the cylinder, and m_1 is cylinder mass.

From equations (1-7) and (1-8)

$$K = \frac{4\pi^2 J_1}{T_1^2 - T_0^2} \tag{1-9}$$

In SI system, the unit of K is $kg \cdot m^2 \cdot s^{-2}$ (or $N \cdot m$).

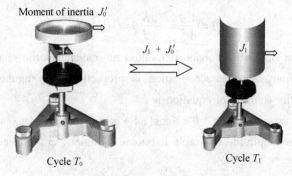

Fig. 1-4 *K* measuring schematic diagram

The parallel-axis theorem of the moment of inertia

As shown in Fig. 1-5, when an object whose mass is m rotates around the centroidal axis O, the moment of inertia of the object is J_0; when the shaft moves parallel distance x, the moment of inertia of the object to new shaft O_1 is $J = J_0 + mx^2$, which is called parallel-axis theorem of the moment of inertia and J is linear to x^2. The concepts of center of gravity and centroid must be distinguished when analyzing the moment of inertia of rigid body.

Center of gravity is the action point of force, while centroid is the center of mass distribution of the object (or system composed of many objects). Different from center of gravity, centroid is not always in the system with gravitational field. There is one thing worth paying attention to: unless gravitational field is uniform, the centroid and center of gravity of the same substance are usually not in the same imagination point. Under normal condition, since the volume of general object is very small compared with that of the earth, the gravitational field, where the

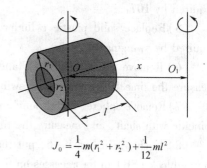

$$J_0 = \frac{1}{4}m(r_1^2 + r_2^2) + \frac{1}{12}ml^2$$

Fig. 1-5 Parallel-axis theorem

object is, is considered uniform, and the centroid coincides with the center of gravity. If the volume of the object is not ignorable compared with that of the earth, the gravitational field, where the object is, is not uniform. Actually the gravitational field is decreased gradually from bottom to top, and the action point of gravity is near bottom, which means the center of gravity is lower than centroid, e.g. the center of gravity of high mountain is a little lower than centroid.

5. Experiment content and operation key points

Experiment contents and steps

The content of this experiment is mainly to measure twist constant of torsion pendulum (twist constant of spring) K; to measure the moment of inertia of solid plastic cylinder, hollow metallic cylinder, solid globe and metallic long and thin rod, and compare it with theoretic calculation value to find out the relative error E_0; verify the parallel-axis theorem of the moment of inertia.

Experiment steps:

(1) Use sliding calipers to measure the diameter of solid plastic cylinder, internal diameter and outside diameter of hollow metallic cylinder, and the diameter of solid sphere, respectively. The length of metallic long and thin rod is known, and record the mass of object to be tested (marked on the object).

(2) Adjust the foot screw of torsion pendulum base to make the bubble in the level be in the middle.

(3) Install metallic carrier disc and adjust the position of photoelectric probe to make the light block bar over the carrier disc in its notch center and block the small hole of transmitting and receiving infrared ray, and measure the time $10T_0$ required by 10 swing cycles.

(4) Put the solid plastic cylinder on the carrier disc vertically, and measure the time

required by $10T_1$.

(5) Replace solid plastic cylinder with hollow metallic cylinder to measure the time $10T_2$ required by swinging 10 cycles.

(6) Remove the carrier disc, and install sphere to the top end of shaft with fixture and measure the time $10T_3$ required by swinging 10 cycles.

(7) Remove sphere, install long and thin rod to shaft to make the center of long and thin rod coincide with shaft, and measure the time $10T_4$ required by swinging 10 times.

(8) As shown in Fig. 1-6, put the sliding block (SB) in the recesses on the both sides of long-thin rod symmetrically, and the distance x between the centroid of sliding block and shaft is 5.00 cm, 10.00 cm, 15.00 cm, 20.00 cm and 25.00 cm, respectively; measure the time of swinging 10 cycles corresponding to different distances. Because the moment of inertia of fixture is very small compared with the moment of inertia of metallic long-thin rod (MLTR), ignore it when calculating.

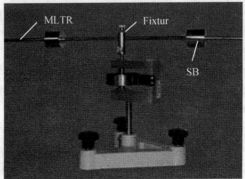

Fig. 1-6 Experiment device diagram for verification of "Parallel-axis theorem of moment of inertia"

(9) Option: put the two sliding blocks asymmetrically, i.e. 5.00 cm and 10.00 cm, 10.00 cm and 15.00 cm, 15.00 cm and 20.00 cm, 20.00 cm and 25.00 cm; and use graphic method to verify parallel-axis theorem of the moment of inertia.

6. Data recording and processing

(1) Measure spring twist constant and the moment of inertia of object, and the data can be seen in Table 1-1.

$$K = \frac{4\pi^2 J_1}{T_1^2 - T_0^2}$$

(2) Verify the parallel-axis theorem of the moment of inertia, and the data can be seen in Table 1-2, where J_5 is the moment of inertia of single sliding rock around the shaft being perpendicular to the center, as shown in Fig. 1-5.

$$J_5 = \frac{1}{16} m_s (D_o^2 + D_i^2) + \frac{1}{12} m_s l^2$$

Where m_s is the mass of the sliding block, D_o and D_i are internal diameter and outside

Measurement of the Moment of Inertia by Torsion Pendulum — Experiment 1

diameter of the sliding block, and l is the length. In this experiment, there is
$$2J_5 = 0.87 \times 10^{-4} \text{ kg} \cdot \text{m}^2$$

Table 1–1(a) The data table for the measurement of the moment of inertia

Object name	Mass /kg	Physical dimension/cm		Cycle /s		Theory value /kg·m²	Test value /kg·m²	Percent difference E_0
Carrier disc				T_0			$J_0' = J_1 \dfrac{\overline{T}_0^2}{\overline{T}_1^2 - \overline{T}_0^2}$	
				$\overline{T}_0$				
Solid plastic cylinder		D		T_1		$J_1 = \dfrac{1}{8}mD^2$	$J_1' = \dfrac{K}{4\pi^2}\overline{T}_1^2 - J_0'$	
		Average value		$\overline{T}_1$				
Hollow metallic cylinder		D_o		T_2		$J_2 = \dfrac{1}{8}m(D_o^2 + D_i^2)$	$J_2' = \dfrac{K}{4\pi^2}\overline{T}_2^2 - J_0'$	
		Average value		$\overline{T}_2$				
		D_i						
		Average value						
Solid sphere		D_{sp}		T_3		$J_3 = \dfrac{1}{10}mD_{sp}^2$	$J_3' = \dfrac{K}{4\pi^2}\overline{T}_3^2$	
		Average value		$\overline{T}_3$				

Table 1 – 1(b)

Object name	Mass /kg	Physical dimension/cm	Cycle /s	Theory value /kg·m²	Test value /kg·m²	Percent difference E_0
Metallic long and thin rod		Average value of length l 61.00	T_4	$J_4 = \dfrac{1}{12}ml^2$	$J'_4 = \dfrac{K}{4\pi^2}\overline{T}_4^2$	
			$\overline{T}_4$			

Table 1 – 2 The data table for verifying the parallel-axis theorem

x/cm	5.00	10.00	15.00	20.00	25.00
T/s					
$\overline{T}$/s					
Theoretical value /kg·m² $J = J_4 + 2mx^2 + 2J_5$					
Test value /kg·m² $J' = \dfrac{K}{4\pi^2}\overline{T}^2$					
Percent difference E_0					

7. Analysis and questions

(1) Why experiment equipment is required to level, why will test error be produced when an object is not set straight?

(2) Why is the mass of fixture not considered when calculating the moment of inertia of solid sphere and metallic long and thin rod?

(3) How to measure the moment of inertia of object in any shape running around specific shaft by using this device?

(4) How to verify parallel-axis theorem by using least square method?

8. The appendix

The principle of measuring the moment of inertia of object by using three-line pendulum

The construction of three-line pendulum is shown in Fig. 1 – 7. The three-line pendulum is, on the circumference of disc, along the apex of equilateral triangle symmetrically connected to the lower regular triangle apex of a bigger uniform disc edge.

When upper and lower discs are in level and three lines are equilong, make the upper disc turn a small angle around vertical central axis $\overline{O_1 O}$ to make big suspended disc do torsion pendulum around central axis $\overline{O_1 O}$ with the help of tension of suspended lines. At the moment, the centroidal O_1 of lower disc will rise and fall along the turning angle, as shown in Fig. 1 – 8. $\overline{O_1 O} = H$ is the vertical distance of centers of upper and lower discs, $\overline{O_1 O_2} = h$ is the rising height of lower disc in vibration, $\overline{OA} = r$ is the radius of upper disc, $\overline{O_1 A_1} = R$ is the radius of lower disc and α is torsion angle.

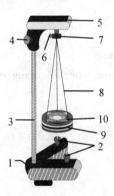
Fig. 1 – 7 Three-line construction diagram
1—base; 2—leveling screws; 3—support;
4—fix screw; 5—suspension; 6—fastening screw;
7—upper disc; 8—suspension line; 9—lower disc;
10—Metallic ring to be measured

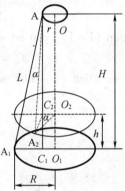
Fig. 1 – 8 Torsional vibration of lower disc

Since the forces of three suspended lines are equal, lower disc motion is symmetric to central axis. We only analyze the motion of one suspended line $\overline{AA_1}$, and L represents the length of suspended line $\overline{AA_1}$, as shown in Fig. 1 – 8. When the lower disc twists an angle α, the suspended line point A_1 of lower disc moves to A_2, the rising height of lower disc is h, the relationship between h and other geometric parameters can be considered as below. Make a vertical line from A of upper disc to lower disc, and it intersects the lower disc before and after rising h with C_1 and C_2, respectively.

In right triangle ΔAC_1A_1:

$$(\overline{AA_1})^2 = (\overline{AC_1})^2 + (\overline{A_1C_1})^2 \tag{1-10}$$

From Fig. 1-8, we can know $AC_1 = OO_1 = H, \overline{A_1C_1} = R - r, \overline{AA_1} = L$, and thus the above equation can be written as

$$L^2 = H^2 + (R - r)^2 \tag{1-11}$$

From $\Delta O_2C_2A_2, \overline{O_2C_2} = AO = r$, so there is

$$(\overline{A_2C_2})^2 = (\overline{A_2O_2})^2 + (\overline{O_2C_2})^2 - 2(\overline{A_2O_2})^2(\overline{O_2C_2})^2\cos\alpha = R^2 + r^2 - 2Rr\cos\alpha \tag{1-12}$$

In right triangle ΔAC_2A_2:

$$(\overline{AA_2})^2 = (\overline{AC_2})^2 + (\overline{A_2C_2})^2 \tag{1-13}$$

Where suppose suspended wire does not extend, and thus there is

$$\overline{AA_2} = \overline{AA_1} = L, \overline{AC_2} = H - h \tag{1-14}$$

So equation (1-13) can be written as:

$$L^2 = (H - h)^2 + R^2 + r^2 - 2Rr\cos\alpha \tag{1-15}$$

Compare equation (1-11) with equation (1-15), and after L^2 being eliminated, we can get

$$H^2 + (R - r)^2 = (H - h)^2 + R^2 + r^2 - 2Rr\cos\alpha \tag{1-16}$$

i. e.

$$h(H - \frac{h}{2}) = Rr(1 - \cos\alpha) \tag{1-17}$$

$\cos\alpha$ is developed to series, there is

$$\cos\alpha = 1 - \frac{\alpha^2}{2!} + \frac{\alpha^4}{4!} - \frac{\alpha^6}{6!} + \cdots \tag{1-18}$$

It is considered that α is little, items after briefing higher than α^2, because h is infinitesimal compared with L and H, microscale which is higher than one step can be eliminated, and obtained from equation (1-17)

$$h = \frac{Rr\alpha^2}{2H} \tag{1-19}$$

When torsion angle α of lower disc is very small, the vibration of lower disc can be regarded as ideal simple harmonic vibration, whose potential energy E_p and kinetic energy E_k are

$$E_p = m_0 g h \tag{1-20}$$

$$E_k = \frac{1}{2} J_0 \left(\frac{d\alpha}{dt}\right)^2 + \frac{1}{2} m_0 \left(\frac{dh}{dt}\right)^2 \tag{1-21}$$

In equation (1-21), m_0 is the mass of lower disc, g is gravity acceleration, $d\alpha/dt = \omega$ is angular frequency, dh/dt is the rising speed of lower disc, and J_0 is the moment of inertia of disc to axis OO_1.

If the effect of friction force is ignored, mechanic energy is conservative in the gravitational field, i. e.

$$\frac{1}{2} J_0 \left(\frac{d\alpha}{dt}\right)^2 + \frac{1}{2} m_0 \left(\frac{dh}{dt}\right)^2 + m_0 g h = \text{constant} \tag{1-22}$$

Since the rotational energy of lower disc is much higher than the translational energy of seesaw motion, i. e. $\frac{1}{2} J_0 (\frac{d\alpha}{dt})^2 \gg \frac{1}{2} m_0 (\frac{dh}{dt})^2$, there is approximately

$$\frac{1}{2} J_0 \left(\frac{d\alpha}{dt}\right)^2 + m_0 g h = \text{constant} \tag{1-23}$$

Substitute equation (1-19) into equation (1-22) and derive t, and then obtain

$$\frac{d^2 \alpha}{d^2 t} = -\frac{m_0 g R r}{J_0 H} \alpha \tag{1-24}$$

Equation (1-24) is a simple harmonic vibration equation, the solution of which is

$$\omega^2 = \frac{m_0 g R r}{J_0 H} \tag{1-25}$$

As vibration cycle $T_0 = -\frac{2\pi}{\omega}$, substitute it into equation (1-25), and then we have

$$\frac{4\pi^2}{T_0^2} = \frac{m_0 g R r}{J_0 H} \tag{1-26}$$

So there is

$$J_0 = \frac{m_0 g R r}{4\pi^2 H} T_0^2 \tag{1-27}$$

It can be seen that the moment of inertia J_0 of lower disc can be precisely calculated as long as the parameters m_0, R, r, H and T_0 which are related to three-line pendulum and can be measured precisely.

To measure the moment of inertia of an object whose mass is m, the moment of inertia J_0 of lower disc without load can be measured first, and then put the object to be measured on the disc, and note that the centroid of object is just on the rotational axis of equipment. Measure the rotational cycle T_1 of the whole system, and thus the moment of inertia J_1 of the system can be calculated through the following equation.

$$J_1 = \frac{(m_0 + m)gRr}{4\pi^2 H_1} T_1^2 \qquad (1-28)$$

Where H_1 is the space between upper and lower discs after the object to be measured is laid; in general, it can be considered as $H_1 \approx H$. The moment of inertia of the object to be measured is J

$$J = J_1 - J_0 = \frac{gRr}{4\pi^2 H}[(m_0 + m)T_1^2 - m_0 T_0^2] \qquad (1-29)$$

With this method, the moment of inertia of objects with any shape can be measured when satisfying the conditions of experiment requirements.

It is known that the moment of inertia of object depends upon the distribution of the shape and mass of the object and the position relative to shaft. Therefore, the moment of inertia of objects varies with different shafts: the shaft can pass through the inside of object, and can also be outside of the object. As for two parallel-axes, the moment of inertia J_α of object to any shaft is equal to the moment of inertia J_c with centroid as axis when passing through the object plus object mass m and the product of distance square d^2 between two axes. This is parallel-axis theorem, whose expression is

$$J_\alpha = J_c + md^2 \qquad (1-30)$$

By changing the distance between the centroid of the object to be measured and the center shaft of three-line pendulum, measuring the relation of J_α and d^2 can verify the parallel-axis theorem of the moment of inertia.

Measurement of Young's Modulus of Metallic Wire by Tension Method

1. Background and application

The shape of solid can be changed by the action of external force, which is called "Deformation". When the external force within certain limit and after the external force action stoping, the deformation completely disappears, which is called "Elastic deformation". When the external force is too high, the deformation can not disappear completely and surplus deformation remains, which is called "Plastic deformation", i. e. that deformation still exists after the external force is removed, which is an irreversible process. Increasing the external force gradually till surplus deformation appears can achieve the object elastic limit.

An important physical quantity describing the capability of solid materials resisting deformation is called Young's modulus, a concept in mechanics of material, put forward by the British physicist Thomas Young (1773—1829) in 1807. It reflects the relation between material elastic deformation and internal stress, and it is the basis to select mechanic component material and is one of the often used parameters in engineering.

Experiment proves Young's modulus is not related to external force applied, object length and sectional area, but it only depends upon solid material structure, chemical component and manufacture method. It is an important physical quantity of solid material to reflect its nature; Young's modulus marks material rigidity: high Young's modulus shows that the deformation of material is low when it is compressed or tensioned. Sound speed in solid is also related to Young's modulus. When Young's modulus is high, sound speed is fast, which is similar to yield coefficient of spring, i. e. the harder the spring is, the faster the wave transmits. The measurement of Young's modulus is very important to study the nature of mechanics nature of various materials such as metallic materials, optical fiber material, semiconductor, nanometer material, polymer, ceramics and rubber, etc. It is also used in fields such as the design of mechanic components, biomechanics and geology, etc.

There are a lot of methods to measure Young's modulus, which can be divided into three classes: The first is static measure method, including static tension, static torsion, and static bending; the second is dynamic measure (resonance measure), including bending resonance (lateral resonance), longitudinal resonance and torsion resonance; and the third is wave velocity measurement, including continuous wave and pulse wave. There are some other methods using advanced techniques, such as inductance displacement measurement, magnetic induction, morse stripe, etc. In the above methods, two methods are mainly used nowadays: One is static tension—optical lever magnification, the basic principle of which is that under the torsion metallic wire, adding weight directly as the way of increasing force, the micro elongation produced is measured by optical lever, and the experiment principle of which is directly from the mechanics definition of Young's modulus, which has a clear visualization; the other is dynamic measurement, i.e. suspension coupling bending resonance method.

A brief introduction to Thomas Young

Thomas Young (1773—1829) (see Fig. 2 - 1), a British physicist, doctor, archaeologist and the great founder of wave optical, made important contributions to opticals, physiological opticals and mechanics of materials, etc.

(1) Wave optical—dual-gap interference

In 1801, Mr. Young published the book, "Essentials of Experiment and Exploration of Sound and Light", systematically discussed the wave viewpoint of light and challenged Newton's "Particle Theory of Light". Young thought that wave theory is more effective than corpuscular theory in explaining why the transmission velocity of bright light and weak light is the same. He also proved that the double refraction phenomenon seen by Huygens in Iceland spar was correct.

Fig. 2 - 1 Thomas Young

In order to confirm the correctness of light wave theory, Thomas Young got two coherent sources with very clever methods, and made the famous experiment of Young's dual-gap interference. His initial experiment method was that use the orifice illuminated by bright light as the point source of light to give out sphere wave; two additional orifices are set at a certain distance from the orifice, which separate the sphere wave given out by the first orifice into two very small parts as the coherent sources; and thus the interference phenomenon was produced in the meeting area where the two orifices gave out light wave, and interference pattern was obtained at the receiving screen behind the dual holes.

(2) Visual optical—trichromatic color principle

Thomas Young applied optical theory in medicine and made the foundation of visual opticals. He put forward the viewpoint that eyes observed different distance object by changing curvature of eyeball crystalline lens for adjustment, which is the earliest explanation of eye opticals principle. He also put forward the assumption that the ability of people to distinguish color was due to some different structures on retina to feel red, green and blue light ray separately, with which the cause of color blindness could be explained. He also established trichromatic color principle, and thought all colors were formed by mixing three primary colors—red, green and blue—to different proportion, which has been the base of modern color theory.

(3) Mechanics of material—Young's modulus

Thomas Young first put forward the concept of modulus of elasticity in mechanics of material, and proposed shear stress was also a kind of elastic deformation. Later his name was used to name modulus of elasticity, called Young's modulus.

(4) Archaeology—the characters on ancient Egypt stone tablet

In 1814, he started to study the ancient Egypt stone tablet found in archaeological studies. Several years later, he translated the characters on the stone tablet, and made outstanding contribution to archaeology.

2. Experiment principles

A uniform metallic wire, whose length was supposed as L and sectional area as S, is elongated ΔL after both ends were pulled by external force F. The experiment shows that within the elastic range of tenable Hook's law, vertical acting force F/S (normal stress) on unit area is proportional to the specific elongation $\Delta L/L$ (linear strain) of the metallic wire, whose proportional coefficient is called Young's modulus. In the system of international units, the unit of Young's modulus is Pa, which is expressed by E, i.e.

$$E = \frac{F/S}{\Delta L/L} = \frac{FL}{S\Delta L} \qquad (2-1)$$

Where F, L and S are all easy to measure, ΔL is micro variable, which is difficult to measure accurately by general methods and here is measured through optical lever magnification method. New digital display hydraulic pressure Young's modulus measuring instrument is introduced to replace the traditional weight forcing, and the space occupied by instrument is reduced, and double reflectors are used to amplify the magnification power.

Magnification method is a widely used measuring technique. In physical experiment, measurement techniques such as mechanic magnifying, optical magnifying, electronic magnifying, etc., will be used. Spiral micrometer enhances measuring precision through mechanic magnification; oscilloscope is used to observe objects by enlarging electronic signals. Optical lever method used in this experiment belongs to optical magnifying technique, the principle of which is widely used in many instruments with a high sensitivity, such as photoelectric reflective galvanometer and ballistic galvanometer, etc. The core of magnification method is to input micro variation into a magnifier, and precise measurement is made again after magnifying.

Suppose micro variation is expressed with ΔL, the magnified measuring value is N, $A = N/\Delta L$ is called magnifying power of the magnifier. In principle, the higher the A is, the easier the measurement will be.

(1) The principle and formula of measuring Young's modulus

Suppose the diameter of metallic wire is d, substitute $S = \dfrac{\pi d^2}{4}$ into equation (2-1), and there is

$$E = \frac{4FL}{\pi d^2 \Delta L} \tag{2-2}$$

(2) The magnification principle of optical lever

A structural scheme of new optical lever is shown in Fig. 2-2. There are pointed screws A, B, and C on each angle of the isosceles triangle, and the two screws B and C on the connected bottom side line are called front foot tips, and the screw A on the apex is called rear foot tip. The optical lever adjuster in the figure can make the reflector turn in level and adjust pitch angle. Measuring scale is on the side of optical lever reflector and is on the same plane with the reflector. When measuring, two front foot tips are placed on the fixed platform of Young's modulus measuring equipment, and the rear foot tip is on the level pallet fixed and connected with the bottom end grip holder of the metallic wire. When the metallic wire is stressed, micro elongation is produced, and the rear foot tip make micro motion with the pallet, and make the optical lever turn a micro angle around front foot tip to drive optical lever reflector to turn a relative micro angle, so the image of measuring scale reflects between optical lever reflector and adjusting reflector, and the micro angular displacement is magnified into bigger linear displacement. This is the basic principle of optical lever producing optical magnification.

Measurement of Young's Modulus of Metallic Wire by Tension Method

Principle formula of measuring elongation in the experiment can be derived below.

The optical lever for the experiment is shown in Fig. 2 – 3: the connected line distance b from rear foot tip A to front foot tips B and C is called constant of optical lever. Fig. 2 – 4 is the magnifying principle scheme of NKY – 2 optical lever. Scale and observer are on both sides; at the beginning, optical lever reflector and scale are on the same plane, and the scale reading on telescope is n_0; When the rear foot tip of optical lever reflector decreases ΔL, a micro deflection angle θ will be produced, and at the moment the scale reading read in telescope is n_1, and $n_1 - n_0$ is elongation N of steel wire magnified, which is called visual elongation. In accordance to the geometric relation of each physical quantity in Fig. 2 – 4, there is

$$\Delta L = b\tan\theta \approx b\theta$$
$$N = n_1 - n_0 = D\tan 4\theta \approx 4D\theta$$

So its magnification power is $A = \dfrac{N}{\Delta L} = \dfrac{n_1 - n_0}{\Delta L} = \dfrac{4D}{b}$

Substitute it into equation (2 – 2), and we have

$$E = \frac{16FLD}{\pi d^2 bN} \tag{2-3}$$

Where D is the distance from adjusted reflecting plane mirror to the scale.

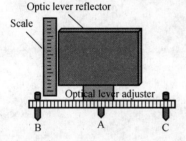

Fig. 2 – 2 New optical lever object drawing

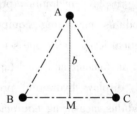

Fig. 2 – 3 Optical lever constant b scheme

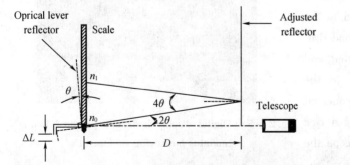

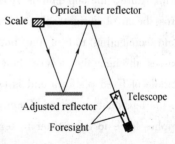

Fig. 2 – 4 Scheme of optical lever magnifying principle

3. Experiment purposes

Young's modulus is an important index to judge material elastic limit and an important concept in mechanics of materials. Through the experiment, first we understand physical significance expressed by Young's modulus and its important function in life. In the process of measuring Young's modulus through elongation method, a key problem is the indirect measurement of micro displacement, and optical lever magnifying method solves the problem cleverly. Through optical reflection principle, the indirect measurement of micro displacement is realized when object micro displacement is magnified by optical lever. It is hopeful in the experiment to learn the process, deeply understand optical magnifying principle, note the effective factors during the experiment and know the significance of processing data through successive differences method. When grasping elongation method, we can know other methods of measuring Young's modulus at the same time.

4. Experiment instruments

Digital display hydraulic pressure Young's modulus measuring equipment, new type optical lever, telescope set, light ruler, spiral micrometer and steel tape.

Digital display hydraulic pressure Young's modulus measuring equipment is shown in Fig. 2 - 5: The top and bottom ends of the metallic wire are fastened by drill fixture, the top end is fixed to the cross beam of upright columns, and drill held connecting rod of the bottom end passes through the sleeve hole in the middle of fixed platform and is connected to the tension transmitter. Forcing device applies force to transmitter to tension the metallic wire.

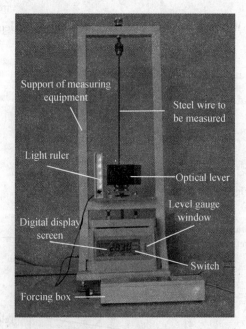

Fig. 2 - 5 Digital display hydraulic pressure Young's modulus measuring equipment

The size of force applied is displayed on liquid crystal display screen and adjusted by hydraulic adjusting valve. Of displayed on the screen indicates that the force applied exceeds the range of measurement or loading when starting the equipment. Solution: Remove the external force or open hydraulic adjusting valve to release some liquid. UF shows the force applied is less than zero point. Solution: Press zero setting key or adjust hydraulic valve to fill in some liquid.

Telescope set is composed of telescope and adjustable reflector, as shown in Fig. 2-6. New type optical lever object diagram is shown in Fig. 2-7. As for the operation method of spiral micrometer, see Experiment 1.

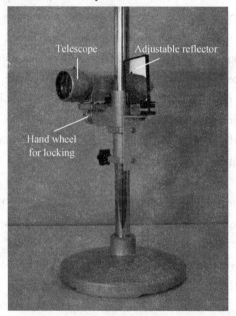

Fig. 2-6   Telescope set

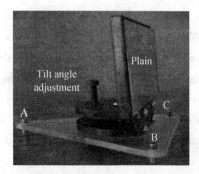

Fig. 2-7   New optical lever object diagram

5. Experiment content and operation key points

(1) Leveling

Observe whether the bubble of round water level on Young's modulus measuring equipment is in the middle; if not, adjust the four foot screws till water level bubble is in the middle, at the moment the columns of Young's modulus equipment are vertical and the platform is in level.

(2) Zero setting

Insert the end of hydraulic connecting pipe into tensile equipment interface, and tighten the compression nut. Make hydraulic adjusting screw stem set to "zero" along force reducing direction (Note: Clockwise turning screw stem is force increasing direction, while counterclockwise turning is force reducing direction).

(3) Adjustment of optical circuit

①Place the optical lever reflector shown in Fig. 2-7 on the platform, the two front foot tips B and C of optical lever on two transparent columns, and the rear foot tip A on the round tray fixed to the steel wire. Adjust tilt angle adjustable screw of optical lever reflector to make optical lever reflector perpendicular to the experiment platform.

②Adjust the focal distance of eye lens of telescope to make it clearly see the scale plate of telescope. Adjust lens cone height of telescope to make it basically the same as the height of optical lever reflector. Move the whole telescope support, and the adjustable reflector is found from optical lever reflector. Along the straight line formed by two foresights from outside of telescope, still observe along the direction and turn the adjustable reflector slightly on both sides till scale image is found. And use eyes to observe from the inside of telescope to see if there is scale; if the scale is not seen, adjust the focal distance of object telescope or adjust micro adjustable screw of reflector slightly till scale graduation line can be clearly seen.

(4) Measurement

①Press "ON/OFF" button of digital-display force gauge; after "0.000" appears on the display, use adjusting bolt of hydraulic forcing box to increase force, and the applied tensile force appears on the screen.

②In order to measure data accurately and conveniently, first measure the loading process: digital display tensile force begins from 14.00 kg. Record scale reading every 1.00 kg, read 10 groups of data, and record them as $n_0, n_1, n_2, n_3, n_4, n_5, n_6, n_7, n_8, n_9$, respectively. A few minutes later, continuously decrease the loading, observe scale reading every 1.00 kg reducing. Read the corresponding 10 groups of data and fill them in the record table.

③Repeat the above steps in (2).

④When the observation is finished, turn hydraulic adjusting screw completely out to make force gauge indicate near "0.000", and then turn off the power switch of the force gauge.

⑤Measure the values of D, L, b and d, whose error limits should be estimated in accordance to specific conditions. D, L, b are only measured once; the method of measuring optical lever constant b is that three foot tips stamped on hard paperboard to make an isosceles triangle and the vertical distance of connected line from rear foot tip to two front foot tips is b. Because of stamping, the width of graphing connecting line can be up to 0.2~0.3 mm, and its error limit estimation is 0.5 mm. When L and D are measured with steel ruler, its limit error can be estimated as 1 ~ 3 mm. For d, use micrometer to measure different positions of metallic wire for six times, average them, it should be paid attention that measuring points must averagely distribute to different positions—top, middle and bottom—and instrument error of micrometer can be taken by 0.005 mm.

(5) Operation points

The adjustment of optical circuit is the base of the experiment, so it is necessary to completely understand the magnifying principle of optical lever (light is transmitted along straight line, and incidence angle is equal to reflection angle). Adjust the optical circuit system of scale—reflection plane mirror—telescope to make the reflection image of scale in plane mirror pass through telescope; adjust focal distance of eye lens and object glass of telescope to make sure that in the telescope, three alignment lines of cross hair plane and scale image graduation line can be seen clearly without parallax. Know clearly the adjusting method of pitch angles of optical lever and adjustable reflector; be gentle, careful and precise when operating.

(6) Attentions

①False elongation will appear if the steel wire is not straight or drill bit fixture is not tight during clipping, so the steel wire must be clipped tightly by drill bit by force. At the same time, before measuring, metallic wire must be straightened and appropriate pretension is applied.

②When the steel wire is applied external force, stable elongation can be achieved after a period of time, which is called hysteresis effect, and the time is called relaxation time. Therefore, after every time the force is applied, don't measure and read data until the data on the display is stable.

③After corroded or long-time stressed, the metallic wire (the steel wire) will appear so-called fatigue of metals, which causes stress concentration or elastic deformation; therefore, when the steel wire is corroded or used over two years, it should be replaced.

④The indicated value error of digital force gauge used in the experiment is ±0.01 kg.

6. Data recording and processing

(1) See scale reading record in Table 2 – 1.

(2) Measure and record D, L and b in the table, and record d data.

(3) Data process and result expression: First use $E = \dfrac{16FLD}{\pi d^2 bN}$ to derive relative uncertainty conduction formula $\Delta E/E$, and then obtain ΔE. N is required to process by successive differences method. When calculating E, force F corresponding to the elongation N is $F = 5.00 \times$

g N, and the unit of E is Pa.

Table 2 – 1   Table of data record

$\Delta F = 0.01$ kg $\Delta_{\text{In}} = \quad$ cm

Frequency	Indicated tensile value/kg	Scale reading/mm				Successive difference value/mm
		Load increase	Load decrease	Average		
0	14.00				$N_1 = \lvert n_5 - n_0 \rvert$	
1	15.00				$N_2 = \lvert n_6 - n_1 \rvert$	
2	16.00				$N_3 = \lvert n_7 - n_2 \rvert$	
3	17.00				$N_4 = \lvert n_8 - n_3 \rvert$	
4	18.00				$N_5 = \lvert n_9 - n_4 \rvert$	
5	19.00				$\overline{N}$	
6	20.00				S_N	
7	21.00				D	_____ $\pm \Delta D$
8	22.00				L	_____ $\pm \Delta L$
9	23.00				b	_____ $\pm \Delta b$

7. Analysis and questions

(1) For Young's modulus measuring data, if graphing method is used to process data rather than successive difference method, how to process?

(2) In accordance with error analysis, if the experiment result of E is made to be ideal, what quantity should be accurately measured, why? Try to exemplify why different length measurements should be measured with different devices?

(3) What is the merit by using optical lever magnification method to measure micro length variation? How to amplify magnification power of optical lever magnification system?

(4) Try to prove: If optical lever reflector is not parallel to adjustable reflector before measuring, measured results can not be affected.

Measurement of Surface Tension Coefficient of Liquid

1. Background and application

Many physical phenomena in life are related to surface tension of the liquid, e. g. the small drops on leaves and grass take on the shape of sphere; when the mercury in broken thermometer drops on the ground, it takes on the shape of sphere; flat things such as coins can keep afloat on the water surface; some insects such as water strider can walk on the water surface. The concept of surface tension was first proposed by Hungarian physicist—Jan Andrej Segner (1704—1777) in 1751, and later many physicists such as Thomas Young in 1805, Pierre Simon Laplace in 1806, Siméon Denis Poisson in 1830 and Joseph Plateau from 1842 to 1868 made great contributions to the theory of surface tension (see Fig. 3 – 1).

Fig. 3 – 1 Surface tension phenomena

Surface tension of the liquid is essentially the macroscopic token of intermolecular interaction. Because there are few molecules in gas phase layer above liquid, the molecules in the surface layer of liquid (the thickness of the surface layer of liquid is the effective radius of the molecules and is about 10^{-7} mm) suffer less upward force than downward one, which results in a resultant force that is perpendicular to liquid surface and pointing to the interior of the liquid; as shown in Fig. 3 – 2, the molecules in the surface layer of liquid has the tendency of inserting into the inner liquid.

Fig. 3 – 2 **The strength performance of molecules in the surface layer of liquid**

Due to the intermolecular interaction of the surface layer of liquid, the surface of liquid is like a piece of intense elastic film, and the surface area of which has a tendency of shrinking into the minimum size. We call the force surface tension that makes the surface of liquid shrink along the surface of liquid. If assuming to scribe a straight line on the surface of liquid, the surface tension represents the interaction of certain tension from the surface beside the line. The tension f exists on the surface layer, whose direction is perpendicular to the straight line permanently and whose size is directly proportional to the length of the straight line L, i.e.

$$f = \alpha L$$

Where proportional coefficient α is called surface tension coefficient of the liquid and the unit is N/m, representing the surface tension of the surface of liquid beside the line in unit tension. The fact indicates that the strength of the surface tension is not only related to the substance that forms the surface, but also changes with the adulteration of the liquid, solution concentration and temperature.

Surface tension is an important characteristic of liquid, and is widely applied in daily life and industrial technologies, such as crystallization, welding, flotation technology, liquid transportation technology, plating technology, cast form, etc.; the surface tension can explain many phenomena involving the surface of liquid in daily life, such as capillarity, formation of foam, spray fogging, etc.; the fluid motion and balance in the bodies of animals and plants, the water motion in the soil, drug preparation and compounding, etc. are also related to the surface phenomenon of liquid.

Since the surface tension coefficient is related to the impurity, i. e. the change of impurity can increase or decrease the surface tension coefficient. When the molten steel is crystallizing, adding a small amount of boron can promote liquid metals to increase the speed of crystallization.

In the process of welding, the surface tension of the solder has a negative influence on welding. When welding, the solder is basically in liquid state, while the pipe legs or pad of the

Measurement of Surface Tension Coefficient of Liquid — Experiment 3

component is solid; when the two substances contact each other, due to the surface tension of liquid, the contact interface of them will decrease, which can influence the area, volume or shape of alloy forming, and surfactant of the flux is needed. Surfactant usually refers to a substance that can obviously reduce the surface tension of other substances with very low concentration. The additive volume of surfactant in flux is very small but quite important: it can guarantee killed spirit smoothly to extend, flow, wet, etc. on the surface of the welded substance.

In the flotation of ores, put ores into the pool filled with water and oil, in which oil is only fully wettable to useful ores and so they are covered by a thin layer of oil; send air into the pool, and thus air bubbles are attached to the useful ores to make them float on the water surface and be separated from impurities such as rocks, etc.

The applications of the surface tension can be seen easily in daily life. The spots on clothes usually have great surface tension and are not water-soluble; laundry powders with surfactant can reduce the surface tension of the spots and make them water-soluble, and thus get rid of the spots from dirty clothes. The surface tension is also very important to the size of the ink drop and the fluid of the ink in the inkjet box and spray nozzle of the printer: ink is in form of fluid when printing, and thus the ink with a low surface tension has better wetting property than that of high surface tension.

Surface tension is also widely applied in biology, medical science and microcirculation system. As we know, many animals including human beings use lungs to breathe. In lungs, there are millions of alveoli (the average diameter of alveolus is about 0.1 mm) to fulfill respiratory function. There are lots of capillaries among alveoli, where the oxygen in the air and the carbon dioxide in the blood are exchanged. The mucus layer in the alveolus is a monolayer covering the internal wall of alveolus, and a kind of surfactant with a function of adjusting the surface tension of the liquid layer of the internal wall of alveolus. The surfactant of the mucus layer of the internal wall of alveolus plays a vital role in adjusting surface tension in breathing process, which can not make the alveoli burst or shrink. Measuring the surface tension coefficient of the liquid in alveoli can confirm the activity of the surfactant and thus relevant diseases can be diagnosed.

There are many ways to measure the surface tension coefficient of the liquid, such as tearing-off method, capillary method, maximum bubble pressure method, drop weight method, etc. In this experiment, the capillary method is used to measure the surface tension coefficient of the liquid.

2. Experiment principles

(1) The surface phenomenon of the interface between liquid and solid

When liquid contacts solid, because of the existence of surface tension on the interface between liquid and solid, the tangents of their surfaces form a certain angle called the contact angle. The contact angle is only determined by the property of the liquid and solid. When the contact angle is an acute angle [$\theta < 90°$, as shown in Fig. 3-3(a)], it is called that the liquid is wetting to the solid; when the contact angle is an obtuse angle [$\theta > 90°$, as shown in Fig. 3-3 (b)], it is called the liquid is non-wetting to the solid; when $\theta = 0$, it is called that the liquid is fully wetting to the solid; when $\theta = \pi$, it is called that the liquid is completely non-wetting to the solid.

When put a drop of mercury on the glass plate, the shape of the drop is always approximately spherical and it can roll along the glass plate without attaching; at this case, we say mercury is non-wetting to the glass plate. When put a drop of water on an oil-free glass plate, the water not only doesn't shrink into the sphere, but also extends outward along the glass surface and attaches the glass to form a thin layer; in this case, we say the water is wetting to the glass.

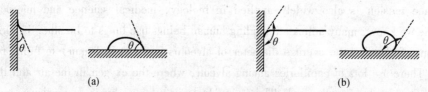

Fig. 3-3 The surface phenomenon of the interface between liquid and solid
(a) Wetting; (b) Non wetting

(2) Capillarity

Because of the surface tension, there is a pressure difference between the inside and outside liquid surface, which is called the additional pressure. In theory, it can be proved that the additional pressure of the spherical liquid with the radius R, is as $P = 2\alpha / R$, where α is the surface tension coefficient. Thus, the larger the surface tension coefficient is, the smaller the spherical radius is and the larger the additional pressure is. For convex liquid surface, the additional pressure is positive, i.e. the pressure inside the liquid is greater than that outside the liquid; for concave liquid surface, the additional pressure is negative, i.e. the pressure inside the liquid is smaller than that outside the liquid, which can be given by $P = -2\alpha/R$.

Measurement of Surface Tension Coefficient of Liquid   Experiment 3

When put a very fine glass tube into the water, you can observe the water height in the tube rises, and the smaller the inner diameter of the tube is, the higher the water will rise. If you put a very fine glass tube into the mercury, the circumstance is contrary, i.e. the mercury height in the tube descends, and the smaller the inner diameter of the tube is, the lower the mercury descends. The phenomenon that the liquid which wets the tube wall rises and the liquid which doesn't wet the tube wall drops is called capillarity. The tube that can be used to generate capillarity is a capillary. This important phenomenon is determined by the surface tension and the contact angle.

As shown in Fig. 3 – 4, when a capillary is just put into the liquid (in this case, the liquid is water or alcohol), since the contact angle is an acute angle (the liquid is wetting to the solid), the liquid level is a concave face, and the additional pressure is negative. The pressure at B below the liquid level is smaller than the atmospheric pressure, and the pressure at C with the same height of B is equal to the atmospheric pressure above the liquid level.

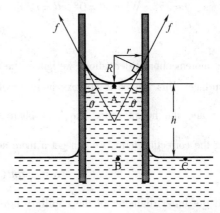

Fig. 3 – 4 **The schematic diagram of using capillary method to measure liquid surface tension coefficient**

According to the basic principle of hydrostatics, when the fluid is static, the pressure of two points with the same height should be equal; therefore, the liquid can't be balanced, and the liquid will rise in the tube until the pressure at B is equal to that at C. The circumstances in convex surfaces are contrary.

(3) The principle of using capillary method to measure liquid surface tension coefficient

Fig. 3 – 4 shows that the glass capillary is inserted into the water. The cross section of the glass capillary, whose inner diameter is r, is circular; the concave liquid surface can approximately be regarded as a sphere whose radius is R; the value of surface tension along tangential direction of concave spherical surface is directly proportional to its circumference $2\pi r$, i.e. $f = \alpha 2\pi r$. The vertical upward force generated by surface tension is $f\cos\theta = \dfrac{2\pi r^2 \alpha}{R}$. This force is balanced with liquid gravity whose height is h, i.e.

$$2\pi r\alpha\cos\theta = \frac{2\pi r^2 \alpha}{R} = \pi r^2 \rho g h \qquad (3-1)$$

Therefore

$$\alpha = \frac{r\rho g h}{2\cos\theta} = \frac{R\rho g h}{2} \qquad (3-2)$$

In this formula, ρ is the liquid density; g is the acceleration of gravity; h is the height from the low end of the inner concave sphere in the capillary to the liquid surface in the container; θ is the contact angle; as for pure water and clean glass, $\theta = 0°$, while as for impure water and general glass, $\theta \approx 25°$. When $\theta = 0°$, $R = r$, the formula becomes:

$$\alpha = \frac{r\rho g h}{2} \qquad (3-3):$$

In the aforementioned derivation, we ignore the liquid quality above the lowest point of the concave spherical surface. This volume is approximately equal to the volume difference between a cylinder with the radius r and the height r and the hemisphere with the radius r, i.e. $\pi r^3 - \frac{2}{3}\pi r^3 = \frac{1}{3}\pi r^3$. Considering the correction term, we can get a more accurate formula than the formula (3-3):

$$\alpha = \frac{1}{2}r\rho g\left(h + \frac{r}{3}\right) \qquad (3-4):$$

When the inner diameter of the capillary is expressed as d, we have

$$\alpha = \frac{1}{4}d\rho g\left(h + \frac{d}{6}\right) \qquad (3-5)$$

Therefore, as long as the inner diameter of the capillary d and the height of liquid h are precisely measured, we can calculate the surface tension coefficient α. The units of each term in the formula are: m for d and h, $kg \cdot m^{-3}$ for ρ, $N \cdot m^{-1}$ for α and g is $9.80 \ N \cdot kg^{-1}$.

3. Experiment purposes

Understand the property of liquid surface; observe the capillarity phenomenon of inserting the glass capillary into the water; understand the principle and method of measuring liquid surface tension coefficient with capillary method; learn how to measure micro-length by using the reading microscope.

4. Experiment instruments

Experiment instruments of measuring liquid surface tension coefficient with capillary method are shown in Fig. 3 – 5, including JCD3 reading microscope, glass capillary, beaker, bracket (used for inserting glass capillary), light source, thermometer, etc.

Microscope is an optical instrument that is used to observe near the wispy objects, the magnification of which is greater than that of the magnifier. Microscope mainly consists of two groups of lenses, as shown in Fig. 3 – 6. One group of lenses facing the object AB is called objective lens; the other group of lenses facing the observer's eyes is called eyepiece. The characteristic of microscope is that the focal length of the objective lens is short while that of the eyepiece is large. The image that is formed through objective lens for the first time is magnified as a real image A_1B_1, and this real image locates within the focal length of the eyepiece and is further magnified into a virtual image A_2B_2 by the eyepiece. The virtual image seen from the eyepiece is magnified twice.

Fig. 3 – 5 Experiment instrument

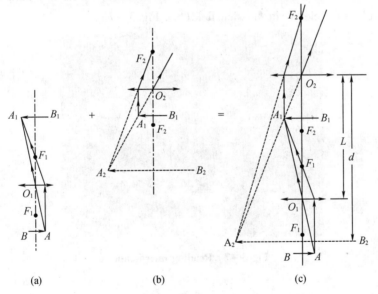

Fig. 3 – 6 The schematic diagram of microscopic imaging

Reading microscope can not only be used to magnify objects, but also to measure the length along a certain direction. It is mainly used to accurately measure tiny objects or objects that can not be measured with a clamp device (such as vernier caliper, spiral micrometer, etc.). JCD3 reading microscope is as shown in Fig. 3 − 5, which mainly consists of microscope, focusing system, the screw micrometer system and working platform. The cross-wire in microscope eyepiece is used to aim at the edge of the measured object to determine the coordinate. Rotating micrometer drum can drive the microscope drawtube to move up and down. Since the pitch of the pushing screw is 1mm, when micrometer drum take a turn, the microscope moves 1 mm from the reading scale. The outer of the drum has 100 scale divisions, and the minimum value is 0.01 mm. It can be estimated to 0.001 mm.

The usage step of reading microscope

(1) Locate the microscope drawtube in the middle of scale;

(2) Use proper light source, and adjust the lighting direction to make the view field bright;

(3) Adjust the eyepiece to make cross-wire clear;

(4) Move the position of the measured object or microscope drawtube to make the optical axis of microscope aim at the measured object;

(5) Adjust the working distance of the microscope drawtube (focusing) until the image with no parallax can be seen clearly in the view field (See Fig. 3 − 7).

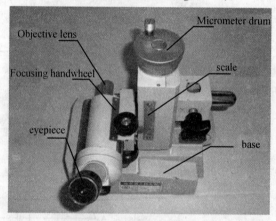

Fig. 3 − 7 Reading microscope

The measuring step of reading microscope

(1) Rotate micrometer drum, drive microscope drawtube to move up and down, make the transverse line of cross-wire at the starting point of the measured object, and write down the position reading a_1 (Read integer on the scale, read decimal fraction on the micrometer drum,

and the sum of the two parts is the reading of this point);

(2) Rotate micrometer drum along the same direction, make the transverse line of cross-wire at the end point of the measured object, write down the position reading a_2, and the distance between the two points is $|a_2 - a_1|$.

The points for attention of using reading microscope

(1) When observing the eyepiece, before focusing the focus knob, first make the objective lens close to the measured object, and then slowly move the drawtube back and avoid the objective lens colliding with the measured object;

(2) For the cross-wires in the objective lens, one should be tangent to the measured object, while the other should be parallel to the moving direction of the drawtube.

(3) Mechanic and reading system are formed by the screw and nut. Because of the gap between the screw and nut, at the beginning of measuring or just reversely rotating the screw, the readings of the drum connected with the screw have some changes, but the instrument connected with the nut does not move. As a result, the reading error occurs, and this error is called a backlash error. To avoid the backlash error, micrometer drum can be only rotated in one direction when measuring; if a problem occurs during the measurement, the measurement must be restarted; forbid to move back and forth during the process, or otherwise the measurement data is meaningless.

5. Experiment content and operation key points

Measuring experiments data include the rising height of water cylinder h, capillary inner diameter d and water temperature T.

Measuring the height of water rising h

(1) Insert the clean glass capillary into the beaker filled with clean distilled water through the hole of the bracket, and pull the capillary up and down to wet the internal wall of the capillary thoroughly.

(2) Pull the capillary upwards slowly, and observe the change of the liquid level in the

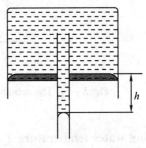

Fig. 3-8　**The schematic diagram of measuring water rising height**

capillary. Measure after the water in the capillary rises to the highest position. When measuring h, the capillary must be upright and there are no bubbles in the water cylinder.

(3) Adjust reading microscope to make the transverse line of the cross-wire at the water surface in the beaker, and write down the position reading h_1; adjust the reading microscope to make the transverse line of the cross-wire tangent to the convex liquid surface in the capillary (the image in the reading microscope is an inverse image), and write down the position reading h_2; then the rising height of the water column is $h = |h_2 - h_1|$. The schematic diagram of measuring water rising height is shown in Fig. 3 – 8.

(4) Repeat the measurements three times, and take the average.

Measuring capillary inner diameter d

(1) Remove the capillary from the water, get rid of the water drops (Notice that you should not blow the water with your mouth), put the capillary on the bracket and make one of its ends aim at the objective lens of the microscope.

(2) Adjust the microscope drawtube to make the transverse line of cross-wire tangent to capillary inner holes, as shown in Fig. 3 – 9. The difference between the two tangent readings is the inner diameter d of the capillary. Turn the capillary, and measure twice in different directions.

(3) Change the other end of the capillary surface to measure twice. Take the average value of the four measurements as the average diameter.

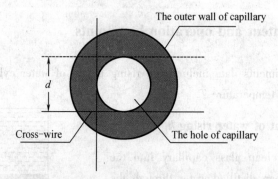

Fig. 3 – 9 The schematic diagram of measuring capillary inner diameter

Measuring water temperature T

Measure water temperature T by the thermometer.

Points for attention

(1) The capillary, beaker and liquid must be clean, or otherwise it will seriously affect the measurement results;

(2) Take care to avoid the backlash error;

(3) Reading microscope is a precise optical instrument; therefore, take care and follow the operation procedures.

6. Data recording and processing

Measuring the rising height of the water h (see Table 3-1)

Table 3-1 Data recording of the rising height of the water h

Unit: mm

degree	1	2	3
h_1			
h_2			
h			
$\bar{h}$			

Measuring capillary inner diameter d (see Table 3-2)

Table 3-2 Data recording of measuring capillary inner diameter d

Unit: mm

degree	1	2	3	4
d_1				
d_2				
d				
$\bar{d}$				

According to the formula (3 − 5), calculate the water's surface tension coefficient α. Compare the surface tension coefficient and the standard value at this temperature, and calculate the relative error. Analyse the main reasons of errors.

7. Analysis and questions

From the experiment principles, capillary method can also be used to measure surface tension coefficient of the mercury. However, mercury cannot wet the glass, and the concave mercury surface in capillary is lower than the surface of the mercury outside the capillary. Since the mercury is not transparent, the distance between the liquid level within the tube and that outside tube can't be measured directly; in addition, the mercury can evaporate in the air and produce toxic mercury vapor; therefore, this method cannot be directly used for measuring the surface tension coefficient of the mercury. In fact, you can cover mercury with water, and then use capillary method to measure surface tension coefficient of the mercury. Think about how to design the experiment to measure and calculate the surface tension coefficient of mercury?

Experiment 4

Determination of Ratio of Specific Heat of Air

1. Background and application

Ratio of specific heat (i. e. specific heat ratio) is a common physical quantity and in thermodynamics theory and applications of engineering and technology play an important role. Such as natural gas transportation safety valve in the process of calculation and the design of the nozzle, we often need to know the specific heat ratio of the gas. When calculating the efficiency of heat engine it also involves the specific heat ratio. In rocket technology, characterization of the energy efficiency of thrust coefficient and velocity are directly related to the size of the specific heat ratio. The transmission characteristic of sound waves in the air is associated with the air specific heat ratio. The determination of gas specific heat ratio in compressor design and experiment also has the extremely vital significance. Specific heat ratio is one of the main parameters of refrigerant. As refrigerant material, specific heat ratio is smaller, and it can reduce exhaust temperature.

Thermodynamics is the study of energy, energy conversion and physical properties related to the energy conversion of the relationship between science. Thermodynamics means heat and power, both produced by thermal power, and reflect the thermodynamics originated from the study of heat engine. Before the 17th century, people had some knowledge and experiences of hot phenomenon, and was widely used in our daily life, but lacked amount of concept and experimental method. The early 18th century, Europe's more developed industry, many production departments such as the development of the steam engine and use, chemical industry, casting involves heat problem, but at the time of temperature and heat the two basic concept confusion, also often see temperature as heat, and hindered the development of thermal. English chemist and physicians and physicist Joseph Black (Joseph Black, 1728—1799, see Fig. 4 - 1) was one of the earliest scientists that carefully studied the nature of heat, he advocated to heat and temperature of two concepts respectively called the amount of "hot" and the strength of the "hot". He found in the heat conduction, with weight and different temperature of the two substances mixed together, the temperature change is not the same,

Fig. 4 - 1   Joseph Black (1728—1799)

he put the matter in the change of the same temperature heat change called these substances "affinity" of heat, "accept heat capacity", and thus puts forward the concept of "heat".

Later his student Irven had introduced the concept of "thermal mass", and carefully measured the specific heat of some materials. Almost at the same time, Swedish scientist also had carried on the research of calorimetry. He pointed out that if the specific heat of water is defined as a unit, it can be mixed with water and other heated object when the temperature changes and the specific heat of the material. For solid and liquid, this difference is very small, generally no longer distinguish, but they are at different temperatures, specific heat will also have change, the same material under different states of matter of the specific heat is different also. In general, the specific heat a substance refers to is the average in a certain temperature range.

For an ideal gas, the equation of state for: $pV = nRT$, the adiabatic process equation for $TV^{(\gamma-1)}$ = constant, $Tp^{(\gamma-1)/\gamma}$ = constant or pV^{γ} = constant. For the constant volume and constant pressure process of gas heating, gas specific heat is constant volume specific heat C_V and specific heat is constant pressure the C_p. Constant volume specific heat is the unit mass of gas heating under the volume unchanged, when the temperature rises 1 ℃ the required quantity of heat, and the specific heat at constant pressure is the unit mass of gas in the case of keeping constant pressure heating, when the temperature rises 1 ℃ heat as needed. The specific heat ratio of constant pressure of the gas C_p and constant volume specific heat C_V is: $\gamma = C_p/C_V$, the coefficient γ is called the specific heat Ratio of the gas (ratio of specific heat capacity), also known as the adiabatic coefficient of gas. According to the first law of thermodynamics, the constant volume gas absorption heat in the process is used to increase its internal energy; And in the process of constant pressure, only one part is used to increase the internal energy of gas, another part is converted to gas against the external force to do the work. So gas increases a certain temperature, the process of isobaric absorbs more heat than the process of constant volume, so the C_p of the gas is larger than C_V, specific heat ratio $\gamma = C_p/C_V > 1$. Generally speaking, C_V and C_p is a function of temperature, but when the actual process involved in the temperature range is not big, both as a constant, and therefore, specific heat ratio also can be regarded as a constant.

The main measuring methods include vibration method, resonance method, sonic velocity method, adiabatic expansion or compression method, etc. Vibration method is that through the implementation process of quasi static in thermodynamics, for the vibration of the vibrating object, measure the cycle to calculate the specific heat ratio. Resonance method is developed on the basis of the vibration method, typical resonance method designed by Ruchar, whose measurement accuracy is higher. Sound velocity method is that the use of sound waves in the ideal gas process can be considered as an adiabatic process, by measuring sound velocity method to calculate the specific heat ratio. This experimental adiabatic expansion method is used to determine the specific heat ratio of air, observation and thermodynamic state changes in the process of the basic laws of physics, deepen the adiabatic, constant volume, constant pressure, the understanding of a few thermodynamics isothermal process. Because the research is not a quasi-static process, the result of the experiment to obtained is rough. But, the experimental method is simple, and help deepen understanding of thermodynamic state in the process of change.

2. Experiment principles

In the air cylinder as the study of the thermal system, experiment indoor temperature T_0, atmospheric pressure for p_0. For with an air inlet and outlet of air cylinder, an open air piston, air cylinder to the atmosphere after close the deflated piston, gas bottle temperature T_0, atmospheric pressure for p_0.

Open the inlet piston, using a balloon pump up from the inlet to the bottle, filling the close rapidly after a certain amount of gas inlet pistons. Due to cheer process quickly, the bottle can't get their gas for heat exchange with the outside world. We can think this is an adiabatic compression process. At this time gas is compressed inside the bottle, while the pressure increases, and the temperature rises.

Gas temperature is higher than the room temperature due to the bottle, the bottle gas through the vessel wall to the outside heat, until at room temperature, stable gas inside the bottle, the bottle of gas in I (p_1, V_1, T_0). That's constant volume heat process.

Then quickly open air piston, connecting the gas inside the bottle and the atmosphere and when the pressure drops to atmospheric pressure p_0 gas inside the bottle, immediately shut off the gas piston. Because the deflated process is faster, reserves of gas bottle too late for heat exchange with the outside world, we can consider this is an adiabatic expansion process. After the process gas in the bottle by state I (p_1, V_1, T_0) into state II (p_0, V_2, T_1). V_2 for air cylinder volume, V_1 is to retain in this part of the gas volume in the bottle in state I (p_1, V_1, T_0).

With gas bottle temperature T_1 below room temperature T_0, so it will slowly from the absorption of heat, gas inside the bottle until at room temperature, the gas bottle pressure also increase as p_2, stable state of gas for III (p_2, V_2, T_0). From the state II (p_0, V_2, T_1) to III (p_2, V_2, T_0) process is a constant volume heat process. The change of system from state I to state III process as shown in Fig. 4-2.

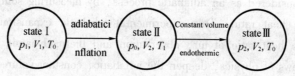

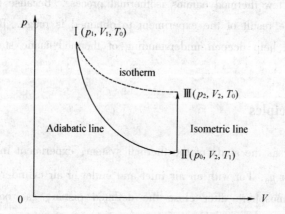

Fig. 4-2 The system state transition graph

State I (p_1, V_1, T_0) to state II (p_0, V_2, T_1) is adiabatic process, the adiabatic process equation:

$$p_1 V_1^\gamma = p_0 V_2^\gamma \qquad (4-1)$$

For the same system, state I (p_1, V_1, T_0) and state III (p_2, V_2, T_0) temperature is the same, satisfying the equation:

$$p_1 V_1 = p_2 V_2 \qquad (4-2)$$

Simultaneous (4-1) and (4-2), elimination V_1, V_2, we can get:

$$\gamma = \frac{\ln p_0 - \ln p_1}{\ln p_2 - \ln p_1} \qquad (4-3)$$

Obviously, as long as we measured three state pressure p_0, p_1, p_2, then we can get the γ value. But it is not really a quasi-static process, as well as poor deflated time control, pressure factors such as measurement error, the measured results are sketchy.

3. Experiment purposes

Through the experiment the adiabatic expansion method is used to test the air specific heat ratio; and observations are the basic physical and thermodynamic state changes in the process of law; to know a certain amount of gas in the process of adiabatic expansion temperature will be lower, the pressure in the process of constant volume increases with temperature rise; be familiar with the piezoresistive pressure sensor and the principle of current type integrated temperature sensor and method of use.

4. Experiment instruments

Experimental instrument is shown as Fig. 4 – 3. The experimental research object is the air in the gas bomb. The pressure and temperature in the bottle are displayed on the tester after they are measured by sensors.

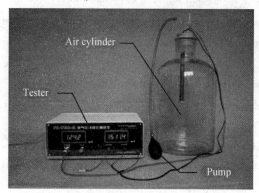

Fig. 4 – 3 Experiment instruments

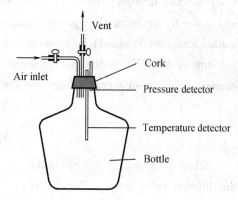

Fig. 4 – 4 Structure chart of gas bomb

The structure of gas bomb is shown as Fig. 4 – 4. Air inlet pipe, an exhaust pipe, an pressure transducer and a temperature sensor are equipped on the bottle stopper. In the experiment, a rubber air sac connected with the air inlet pipe is used to pump to the bottle.

The tester panel is shown as Fig. 4 – 5. The signal line of pressure transducer should be connected with the special interface of air pressure. A three and a half digital voltmeter is used to display air pressure (20 mV/kPa), and the reading is the different value between atmosphere pressure and air pressure in the bottle. The signal line of the temperature sensor should be connected with the special interface of temperature after it is cascaded with 6V DC power supply. A four and a half digital voltmeter is used to display temperature (5 mV/℃).

Fig. 4 – 5 Tester panel

A high precision diffusion silicon piezoresistive differential pressure sensor is used to measure air pressure in the bottle in the experiment. It uses the piezoresistive effect of semiconductor resistance as a sensing principle. In order to improve the sensitivity and stability of the sensor, the bridge circuit composed of the sensing resistor, the signal amplification and processing circuitry integrated on a silicon wafer through diffusion process. As shown

Fig. 4 – 6 Pressure sensor
1—power input(+);2—signal input(+);
3—power input(−);4—signal input(−)

in Fig. 4 – 6, endpoint C of pressure transducer should be connected with measured gas in the bottle. Endpoint D should be connected with atmosphere. Provide a constant input voltage to pressure sensor. When the gas pressure in the bottle changes, sensor output voltage value changes. The sensor output voltage and the pressure change into a linear relationship.

$$U_i = U_0 + K_p(p_i - p_0) \qquad (4-4)$$

$$K_p = \frac{U_m - U_0}{p_0} \qquad (4-5)$$

Where p_i is measured air pressure; p_0 is atmosphere pressure; U_i is the output voltage when the different value between C and D is $(p_i - p_0)$. U_0 is the output voltage of the sensor when the different voltage value between C and D is 0; U_m is the output of the sensor when the different voltage value between C and D is p_0. So, pressure of measured gas can be calculated through following formula:

$$p_i = p_0 + \frac{U_i - U_0}{K_p} \qquad (4-6)$$

Determination of Ratio of Specific Heat of Air Experiment 4

The silicon pressure sensor which is used in this experiment has been demarcated by manufacturers. When measured gas pressure is $p_0 + 10$ kPa, digital voltmeter display 200 mV, and measuring sensitivity is 20 mV/kPa. Measuring sensitivity of each silicon pressure sensor is different. So each pressure sensor has its own tester.

Semiconductor integrated temperature sensors are used to measure temperature. There are a number of commonly used temperature sensors, such as wire resistance temperature sensors, thermocouples, thermistor, infrared radiation temperature measuring devices and semiconductor integrated temperature sensors. Semiconductor integrated temperature sensors are used in this experiment. Semiconductor PN junction feature is used to measure temperature. The AD590 temperature sensor which is used in this experiment is a typical type of current output integrated circuit temperature sensors. It has a high measuring sensitivity, good linear, and its measurement error is very small after calibration. The AD590 temperature sensor is connected with 6V DC power supply, and its measuring sensitivity is 1 μA / ℃. If it is cascaded with 5 kΩ resistance, AD590 can produce 5 mV/℃ voltage signal, and measuring circuit is showed as Fig. 4 −7.

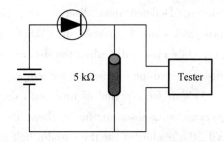

Fig. 4 −7 The temperature sensor wiring diagram

5. Experiment content and operation key points

Preparatory work

(1) Turn on the power of the tester. Preheat the instrument 20 minutes.

(2) Open air inlet valve and exhaust valve. Set three and a half digital voltmeter to zero through adjusting zero setting rotary knob on the panel. Measure indoor temperature T_0 by four and a half digital voltmeter. Atmosphere pressure p_0 is shown by laboratory.

(3) Check the system for leaks. Close exhaust valve of gas bomb, open air inlet valve, and pump up to the bottle. Make the air pressure to 1 ~ 2 kPa (Corresponding voltage value is 20 ~ 40 mV), and then close air inlet valve. Observe pressure value. If the value decline all the time, the system is flat. Bottle stopper and the interfaces on it should be checked. Seal the place where it may be flat with sealant.

Measurement

(1) Open exhaust valve to make gas bomb to connect with atmosphere. The temperature and pressure of the gas in the bottle are same with atmosphere at this time. Close exhaust valve. Open air inlet valve, and pump up to the bottle. When the reading of three and a half digital voltmeter is among 140～160 mV, close the air inlet valve and top to pump up rapidly. Observe the change of the pressure and temperature in the bottle. Wait for a period of time. When the temperature of the gas in the bottle is same with indoor temperature, record Δp_1 (mV), the reading of three and a half digital voltmeter and calculate p_1, the pressure of the gas in the bottle. The gas is state Ⅰ at this time.

(2) Suddenly open air valve. When the air pressure in the bottle decline to atmosphere pressure p_0 (Deflated voice disappears), close the exhaust valve rapidly. The state of the gas become from state Ⅰ to state Ⅱ. When operating, deflated quickly, deflated time should be proper. Only close valve when the deflated voice disappears, earlier or later will take errors into results. Observe the changes of the temperature and pressure of the gas in the bottle.

(3) Wait for a period of time until the gas temperature in the bottle reaches to the indoor temperature after absorbing heat. Then, the state become from state Ⅱ to state Ⅲ. Record Δp_2 (mV), the reading of the three and a half digital voltmeter, and calculate p_2, the air pressure in the bottle.

(4) According to the above steps, operate from state Ⅰ to Ⅱ, and then from Ⅱ to Ⅲ. Observe the changes of temperature and pressure. Calculate air specific heat ratio γ.

(5) Repeat steps 1, 2, 3, 4. Measure five times.

Announcements

(1) Different diffusion silicon pressure sensors have different sensitivity, so please don't borrow tester or gas bomb from each other.

(2) The gas bomb, air inlet valve, exhaust valve and connecting pipe used in the experiment are all made of glass, and they are fragile. So closing or opening valves in the experiment and pumping up to the bottle, all should be careful.

(3) Thermal experiment is strongly influenced by the external environment, especially temperature, when measured, more attention should be paid to the change of environmental temperature. When you are measuring, you should do as that the gas temperature in the bottle decline to a certain temperature, and it can return the same temperature after deflated. The temperature isn't necessarily equal to the indoor temperature before measurement.

6. Data recording and processing

Design data table, record $p_0, \Delta p_1, p_1, \Delta p_2, p_2, \gamma$. Calculate the average of special heat ratio, and compare the result with the standard value $\gamma_0 = 1.402$, and then calculate relative error.

7. Analysis and questions

(1) What are the factors of affecting the γ?
(2) We known that wind from southeast blows in summer and wind from northwest blows in winter. What is the cause of this phenomenon according to the content of this experiment?

8. The appendix

The main measurement methods of gas special heat ratio are adiabatic expansion method and vibration method. The vibration method calculates gas specific heat ratio through measuring period of vibration of the object in a certain container. Experimental basic unit is shown in Fig. 4-8. The diameter of vibration object, iron pellet, is just smaller than the diameter of the glass tube only 0.01~0.02 mm. The iron pellet can move up and down in the glass tube. There is a air inlet opening on the glass wall and a slim tube can be inserted. Measured gas can be injected through the slim tube.

The quality of the iron pellet is m, and radius is r, diameter is d. When the air pressure satisfy the following conditions, the ball is in the state of balance.

$$p = p_0 + \frac{mg}{\pi r^2} \quad (4-7)$$

Where p_0 is atmosphere pressure. Inject small air pressure airflow through air inlet opening, and cut a small opening in the middle of the glass tube. When the vibration object is at the bottom of vibration period under the small opening, inject gas to make air pressure in the bottle increase, so the object move up. And when the object is at top of vibration

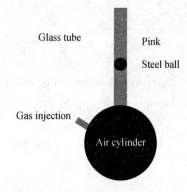

Fig. 4-8 Experiment instruments

period on the small opening, the gas in the container will discharge through the small opening, so the object moves down. Repeat the above process. As long as injection gas flow is controlled properly, the object can do simple harmonic vibration in the glass tube. Vibration period can be measured by the photoelectric timing device.

If the object deviates from the equilibrium position a smaller distance x, then the air pressure in the container changes Δp. The equation of motion of the object is

$$m \frac{d^2 x}{dt^2} = \pi r^2 \Delta p \qquad (4-8)$$

The object vibration is very fast, so changes of the gas state in the container can be seen as adiabatic process, so

$$pV^\gamma = \text{Constant} \qquad (4-9)$$

The derivative of the above formula is

$$\Delta p = -\frac{p\gamma \Delta V}{V} \qquad (4-10)$$

The change of gas volume in the container is

$$\Delta V = \pi r^2 x \qquad (4-11)$$

Combine formula (4-10), (4-11) and (4-8):

$$\frac{d^2 x}{dt^2} + \frac{\pi^2 r^4 p\gamma}{mV} x = 0 \qquad (4-12)$$

The above formula is the differential equation of simple harmonic vibration, its solution is

$$\omega = \sqrt{\frac{\pi^2 r^4 p\gamma}{mV}} = \frac{2\pi}{T} \qquad (4-13)$$

So

$$\gamma = \frac{4mV}{T^2 p r^4} = \frac{64mV}{T^2 p d^4} \qquad (4-14)$$

Here every volume can be conveniently measured, so measured gas specific heat ratio can be calculated.

Experiment 5

Mapping of Electrostatic Field by Imitative Method

1. Background and application

When seeking material motion law and solving engineering problems, people often meet some special cases such as buildings damaged by earthquake and phenomenon of weightlessness in the capsule, etc., which are difficult to research by using direct experiment method. People often artificially create a physical phenomenon or process model similar to the object researched according to similarity theory, and research the phenomenon through models. The method is called simulation method, which can be divided into two great kinds: physical simulation and mathematic simulation.

(1) Physical simulation

The artificial "Model" has the physical process and geometry similar to the actual "Prototype", the analog method based on which is called physical simulation. For example, in order to research the stress of each position of aircraft flying with high speed, people first made a model similar to the prototype in geometry, and then the model was placed in a wind tunnel to create a physical process completely similar to an actual aircraft flying in the air. Through testing stress condition of the model airplane, people can obtain reliable experiment data by using shorter time, convenient space and less cost, as shown in Fig. 5 – 1. Physical simulation has vivid objectiveness and

Fig. 5 – 1   A400M receiving wind tunnel test

makes the phenomenon observed appear repeatedly, so it has widely applied value, especially in researching objects that are difficult to describe accurately by using mathematic equation.

(2) Mathematic analog

The model and the prototype follow the same mathematic law, i.e. satisfy similar mathematic equation and boundary conditions, but their physical nature can have nothing in common. This analog method is called digital-analogy, also called analogue.

Though simulation method has the above merits, it also has a great limitation, and can not increase experiment precision because it can only solve detected problems.

2. Experiment principles

Electrostatic field is determined by electric charge distribution. For some simpler cases, such as spherical conductor, parallel plane board, etc. electric field distribution can be obtained through theoretical calculation. But in most cases, the shape of charged body is more complex, and it is very difficult or has no way to obtain its analytic solution of electrostatic field distribution. Nowadays, computer numerical calculation method can be used to obtain its numerical solution of electrostatic field distribution condition, but the reliability of calculated results still needs verifying, so it is a main method to research electrostatic field distribution characteristic through experiment method. There is a great difficulty in measuring electrostatic field directly. First, there is no current in electrostatic field, so it is not possible to measure it by using simple instrument and the instrument and equipment to be used are very complex; second, once the probe is placed in electrostatic field, induced charge will be produced to make original electric field produce distortion, which affects the accuracy of measured results.

(1) Simulating electrostatic field by using stable and constant electric current field

In order to overcome the difficulty of measuring electrostatic field directly, a stable and constant electric current field can be coined the same as the distribution of electrostatic field to be measured. Use the stable and constant electric current field easy and direct measuring to simulate electrostatic field.

Stable and constant electric current field and electrostatic field are two different fields, but the two fields have similar features. They all have source fields (conservative field), and can all introduce electric potential U to them. For electrostatic field and stable and constant current field, two groups of corresponding physical quantity can be used to describe them, which follow physical as shown in Table 5-1.

From Table 5-1, mathematic forms to describe the physical law of two fields are the same.

Based on the theory of electrodynamics, it can be proved seriously that as for the same equations with the same boundary conditions, and the solution forms are also the same (at most a constant differs), which is the base to use stable and constant electric current field to simulate electrostatic field.

Table 5-1 Physical quantity of describing electrostatic field and stable and constant electric current field

Electrostatic field	Stable and constant electric current field
In the uniform dielectric, each flat plate of two conductors has electric charge $\pm Q$	Current I flows through uniform conducting medium between two electric poles
Potential distribution V	Potential distribution V
Electric field intensity E	Electric field intensity E
Medium dielectric constant ε	Conducting medium conductivity σ
Electric displacement vector $D = \varepsilon E$	Current density vector $J = \sigma E$
When there is no free charge in medium	When there is no power in conducting medium
$\oint D \cdot dS = 0$	$\oint J \cdot dS = 0$
$\dfrac{\partial^2 V}{\partial x^2} + \dfrac{\partial^2 V}{\partial y^2} + \dfrac{\partial^2 V}{\partial z^2} = 0$	$\dfrac{\partial^2 V}{\partial x^2} + \dfrac{\partial^2 V}{\partial y^2} + \dfrac{\partial^2 V}{\partial z^2} = 0$

In order to realize simulation in experiment, boundary conditions of stable and constant electric field and electrostatic field simulated must be the same or similar, which requires that good conductors with the same form and location are used to simulate live conductors generating electrostatic field in simulation experiment, as shown in Fig. 5-2.

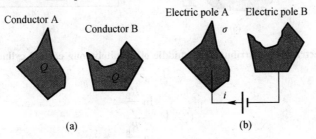

Fig. 5-2 **Comparison of electrostatic field and stable and constant current field**

Because the quantity of electricity on live conductor in electrostatic field is constant and the voltage between two poles of corresponding simulating electric current field is also constant. The conducting medium (bad conductor) in electric current field is used to simulate dielectric in electrostatic field. If the electrostatic field in vacuum (air) is simulated, the conducting medium in electric current field must be uniform medium, i. e. conductivity must be equal everywhere. Because the surface of live conductor in electrostatic field is equipotential surface and the field intensity (or power line) near conductor surface is perpendicular to the surface, what requires the surface of electric pole (good conductor) in electric current is also equipotential. This can be ensured only when conductivity of electric pole (good conductor) is much higher than that of conducting medium (bad conductor); therefore, the conductivity of conducting medium mustn't be over high.

(2) Electrostatic field distribution of the middle of infinitely long coaxial cylinder conductor

As shown in Fig. 5 – 3(a), in vacuum there is an infinitely long cylinder A and infinitely long cylinder shell B placed coaxially (both are conductors), having equivalent contrary sign charge. It is known from electrostatics that the electrostatic field produced between A and B, equipotential surface is a series of coaxial cylinder surfaces and the power lines are some direct lines along radial distribution. Fig. 5 – 3(b) is a schematic diagram of circle equivalent lines in any section S perpendicular to axis and radial power lines distribution. It is known from theoretical calculation that the potential of a point which is away from the axis with the distance of r is

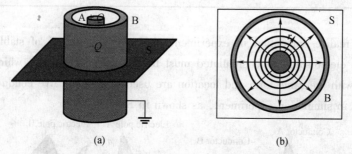

Fig. 5 – 3 Electric field distribution of middle of infinitely long coaxial cylinder conductor

$$V_r = V_1 \frac{\ln \frac{R_B}{r}}{\ln \frac{R_B}{R_A}} \quad (5-1)$$

Where V_1 is the potential of conductor A; the potential of conductor B is zero (to the ground). The field intensity to the center r is

$$E_r = -\frac{dV_r}{dr} = \frac{V_1}{\ln\dfrac{R_B}{R_A}} \cdot \frac{1}{r} \tag{5-2}$$

Where negative sign expresses that field intensity direction points to the potential drop direction.

(3) Analog current field distribution

Fill the conducting medium whose conductivity is very low in the middle of infinitely long coaxial cylinder; add voltage V_1 between inner and outer cylinders; make outer cylinder body grounded to make it have zero potential, at the moment the current passing through the conducting medium is stable and constant current. The current field in conducting medium can be considered as the above simulating field of electrostatic field, as shown in Fig. 5 - 4.

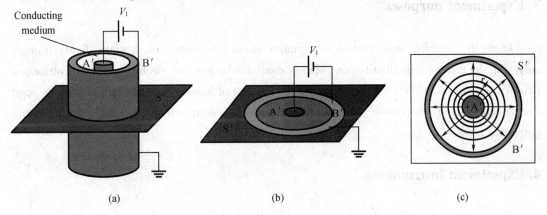

Fig. 5 - 4 Electric field distribution of the middle of infinitely long coaxial cylinder conductor

Since the power line of infinitely long coaxial cylinder is in the plane perpendicular to the cylinder, the power line of simulating electric current field is also in the same plane and its distribution is not related to the position of the axis. Therefore, electric field problem of three dimensional space can be simplified into two dimensional plane problems, i. e. it is OK only to research the electric current line distribution of a conducting medium on a plane.

Theoretical calculation can prove that the potential distribution V'_r of the plane S' in electric current field is completely the same as the potential distribution V_r of power line plane S of electrostatic field in original vacuum, and electric field intensity E'_r in conducting medium is also completely the same as electric field intensity E_r of electrostatic field in original vacuum, i. e.

$$V'_r = V_1 \frac{\ln \frac{R_B}{r}}{\ln \frac{R_B}{R_A}} = V_r \qquad (5-3)$$

Then

$$E'_r = -\frac{dV_r}{dr} = \frac{V_1}{\ln \frac{R_B}{R_A}} \cdot \frac{1}{r} = E_r \qquad (5-4)$$

It can be seen from the above analysis that the distribution functions of V_r, V'_r, E_r and E'_r are completely the same.

3. Experiment purposes

Learn the principle and method of imitative method to survey electrostatic field; through surveying the electric field distribution between coaxial cylinders and electrostatic field distribution between two parallel conductors, understand the concept of imitative method and conditions, and deeply understand the basic physical concepts of electric field intensity and electric potential, etc.

4. Experiment instruments

Experiment instruments: Devices required for the experiment are shown in Fig. 5-5: Power (HY1711-3s Two-way output trackable DC regulated power) and digital avometer, the using method of which can refer to the preliminary knowledge of electromagnetics.

Electrostatic field simulation device 1 is shown in Fig. 5-6, which is used for simulating the electric field distribution in infinitely long live coaxial cylinder conductor. On the conductive glass base carved with coordinate, the middle is a circle electric pole with the radius of 1.00 cm, and the surrounding is a concentric circle ring electric pole with the internal diameter of 10.00 cm. When using the equipment, a voltmeter can be directly used on the equipment surface. Electric pole is a layer of plating with a good conductivity. Don't use detecting probe to touch electric pole when using the equipment, or otherwise peeling off occurs easily to damage the equipment.

Electrostatic field simulation device 2 is shown in Fig. 5-7, which is used for simulating electric field distribution between two parallel conductors. Different from device 1, there are three wires outlet derived by conductor connecting plugs, marked by A, B and D, respectively, among which by A and B are two electric poles and D is a measuring end.

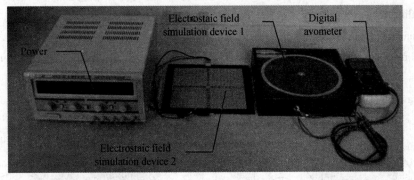

Fig. 5 – 5 Experiment devices

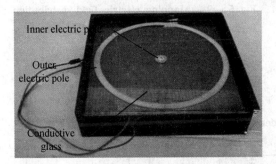

Fig. 5 – 6 The simulation device of coaxial cylinder electrostatic field

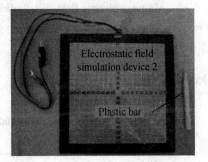

Fig. 5 – 7 The simulation device of parallel conductor electrostatic field

5. Experiment content and operation key points

(1) Mapping equipotential line between coaxial cylinders and drawing power line

Connect wires as shown in Fig. 5 – 8. Experiment steps are as follows:

①Calibrate power voltage (7 V).

②Measure the potential required in Table 5 – 2, and each potential needs uniform measurement of at least 8 points.

③Draw electric field profile according to plotting requirements.

④Measure radius of each equivalent line, and fill in data form and calculate (pay attention to the effective numbers).

(2) Mapping electric field distribution between two parallel conductors

Connect wires as shown in Fig. 5 – 9. Experiment steps are as follows:

①Calibrate power voltage (7 V).

②Measure potential values of conductive plate surface at the positions of 0, 30, 60, − 30, − 60 cm, respectively and mark them on the coordinate paper, and on each equivalent line there are no less than 8 uniform measurement points.

③Draw electric field profile according to plotting requirements.

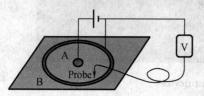

Fig. 5 – 8 Mapping electric field distribution wiring diagram between coaxial cylinders

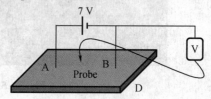

Fig. 5 – 9 Mapping electric field distribution wiring diagram between two parallel conductors

6. Data recording and processing

Table 5 – 2 Coaxial cylinder electric field distribution

($V_1 = 7.00$ V)

V_{rExp}/V	5.00	4.00	3.00	2.00	1.00
r/cm					
$\ln(R_B/r)$					
V_{rTheo}/V					
$E_r = (V_{rThwo} - V_{rExp})/V_{rThwo}$					

$\overline{R_A} = 1.00$ cm $\overline{R_B} = 10.00$ cm $V_{rTheo} = \dfrac{V_1}{\ln\dfrac{R_B}{R_A}} \cdot \ln\dfrac{R_B}{r}$

7. Analysis and questions

(1) Why can stable and constant electrical field be used to simulate electrostatic field, and what are the simulation conditions?

(2) Is it possible to calculate electric field of a point of them in accordance with the described

equivalent line cluster, and why?

(3) If the voltage of power used in experiment is doubled or reduced by half, are the forms of measured equivalent line and electric line of force changed or not?

(4) Calculate their electric field intensity based on the potential values for each position on AB connected line to describe the distribution law of electric field intensity in different areas.

(5) Take 4 ~ 5 points on one of equivalent lines arbitrarily to calculate the ratio of distance from each point to two electric poles and verify if the ratio value is a constant.

8. Appendix

Application of electrostatics

For electrostatic application, people have designed many kinds of processing technology and processing equipments according to the effect and principle of electrostatic induction, gas discharge of high voltage electrostatic field, etc. Electrostatics is widely used in electric power, machinery, light industry, textile, aerospace as well as high technology fields.

Electrostatic precipitation (see Fig. 5 – 10) means to remove micro dust floating in aero by electric method. Dust collecting electric pole is grounded, and discharged electric pole is applied with DC voltage (-40 ~ -200 kV) to form corona discharge.

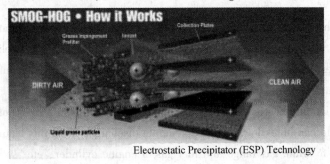

Electrostatic Precipitator (ESP) Technology

Fig. 5 – 10 Schematic diagram of electrostatic precipitation

Dust-laden aero enters into discharge area below dust collecting:
(1) Electric pole and the dust will have negative charge.
(2) Negative charge dust is absorbed by dust collecting electric.
(3) Pole under the action of electric field, so powder dust.
(4) Airflow can be removed.

(5) Electrostatic painting (see Fig. 5 – 11). Make paint micronized and make it have negative charge. The metallic object to be painted is grounded, and the sprayed particles will move along electric line of force to make paint firmly adhesive on the surfaces of objects.

(6) Electrostatic sorting (see Fig. 5 – 12) means to utilize electrostatic force to separate different compositions from the mixture composed of two kinds of particles with different conductivity. For example, place mixed particles on the metal plate, utilize corona discharge to make particles charged. Tilt the metal plate, at the moment since particles with good conductivity lose more charges and the adhesion to metal plate drops, they drop off quickly from the metal plate, and thus realize.

Fig. 5 – 11 YSP series full-auto electrostatic painting production line

Fig. 5 – 12 Electrostatic sorting unit

(7) DC high voltage generation. Van de Graaff electrostatic generator utilizes corona discharge to make high voltage ball charged to generate DC high voltage, which can be used to do insulating check test such as leak current test, etc. and widely used in manufacturers producing power equipment and departments such as power system, etc.

(8) The ignition of fuel gas. Electrostatic has the characteristic of high potential, whose high voltage electric field can breakdown gas and generate discharge. With this feature, design a micro current discharger to ignite gas range and fuel gas of engine cylinder, etc.

Electrostatic hazard

While electrostatics is widely used, electrostatic hazard often appears. The first Apollo manned spacecraft fired and exploded resulting from electrostatic discharge, and three astronauts died; 45% of non-conformed devices in IC production in Japan are caused by electrostatics; in 1988, 5 billion USD lost because of the influence of electrostatic discharge in the USA; in the early 1990s, high-grade digital avometer trial produced by a company in Beijing were not conformed mostly because electrostatics was not prevented by IC. There are a lot of methods to produce electrostatic, such as touch, friction, rush current, piezoelectricity, differential temperature, freezing and electrolysis, etc. Electrostatic prevention is an arduous task and there is a long way to go.

Experiment 6

Measurement of Solenoidal Magnetic Field by Hall Effect Method

1. Background and application

In 1879, 24-year-old Hall (Edwin Herbert Hall, 1855—1938) (see Fig. 6 – 1), was a postgraduate in Johns Hopkins University in Maryland. Under the guidance of his supervisor, Professor H. A. Rowland, he verified whether the magnetic field affected the current in conductor through experiment but found a special phenomenon: As shown in Fig. 6 – 2, put a current—carrying a conductor board into the magnetic field to make magnetic field direction perpendicular to the current direction, and there is cross potential difference U_H between 3 and 4 of the conductor board, which was called Hall effect, and the potential difference U_H formed on both sides of the conductor board was called Hall voltage. At that time, Hall's discovery shook the scientific community, called "The most important discovery in electricity in the past 50 years", and it was praised by Kelvin as the discovery which could compare with Faraday's. But the wide application of Hall effect started about 70 years after it was discovered when semiconductor technique arose.

Fig. 6 – 1 E. H. Hall

Hall effect is widely used in many fields in today's scientific techniques, such as measurement technique, electronic technique, and automation technique, etc. In recent years, new semiconductor materials and the development of cryophysics make people achieve breakthrough progress in studying Hall effect. German physicist K. V. Klitzing won the Nobel Prize for Physics of 1985 for the discovery of quantum Hall effect; Chinese American physicist Cui

Qi, German American physicist H. L. Stormer, and American physicist R. B. Laughlin won the Nobel Prize for Physics of 1998 for a brilliant contribution to the discovery of fractional quantum Hall effect. Cui Qi also became the sixth Chinese American scientist who won the Nobel Prize. This field caused people's wide interest for the awarding of two Nobel Prizes.

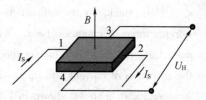

Fig. 6 – 2 Hall effect

Based on Hall effect, people made Hall components with semi-conductor materials, which have the merits of such as being sensitive to the magnetic field, simple structure, little volume, wide frequency response, high output voltage change and long operational life, etc. So they are widely used in the fields of measurement, automation, computer and information technique, etc.

The first commercial Hall device came out in 1959. In 1960, about one hundred kinds of general measurement instruments were developed, with measurement range from 10^{-7} T of constant magnetic field or high frequency magnetic field, which had the advantages of convenient operation, high precision, and were especially suitable to measure the space of little gap.

Hall effect also can be used to measure parameters of semiconductor materials, such as carriers density, conductivity, mobility, to judge the conductive types of semiconductor materials, etc., and to measure non-electricity quantity such as current, force, temperature, displacement, pressure, angle, vibration, speed, and acceleration, etc..

Using plasma Hall effect to manufacture hydro-magneto generators may be a direction to substitute steam-electric power in the future, the basic principle of which is using plasma Hall effect to make positive and negative charged particles of plasma passing magnetic field separated and then gather at two polar plates to form power potential by the action of lateral magnetic field. The new highly effective generation type makes gas become plasma flow and change it into power energy through the heat energy generated by fuel burning. It is not necessary, as steam-electric power does, to first make heat energy released by fuel burning change into mechanic energy drive generator wheel to rotate, and then to change mechanic energy into power energy. Therefore, while the usage efficiency of heat energy is increased, requirements for environment protection are met. At present, there have been example projects for this field, the development prospect of which is broad.

2. Experiment principles

(1) The principle of measuring magnetic field by using Hall effect method

Hall effect, in its nature, is a deflection caused by movable charged particles acted by Lorentz force in magnetic field. When charged particles (electron or hole) are bound in solid materials, the deflection causes the accumulation of positive and negative charges perpendicular to the direction of current and magnetic field, and thus forms additional cross electric field. For the semiconductor material shown in Fig. 6 – 3, if current I_S flows in X direction and magnetic field B is added in Y direction, contrary sign charges start to accumulate in Z direction (on the sides of 3 and 4) to form corresponding additional electric field. The electric field stops the carriers from displacing continuously to the sides. When the electric field force F_e received by carriers is equal to Lorentz force F_m, the accumulation of charges on both sides of the semiconductor will balance, and at the time there is

$$qvB = qE_H \qquad (6-1)$$

Where q is charge electric quantity, E_H is Hall electric field intensity, v is drift velocity of carriers in current direction, and B is magnetic strength of applied magnetic field.

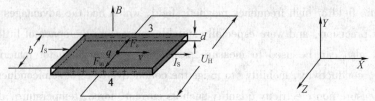

Fig. 6 – 3 The schematic diagram of Hall effect

Suppose the width of Hall material is b, thickness is d, carrier consistency is n and average speed is $\bar{v}$. Then

$$I_S = qn\bar{v}bd \qquad (6-2)$$

It can be obtained from formulas (6 – 1) and (6 – 2) that

$$U_H = E_H b = \frac{1}{nq} \cdot \frac{I_S B}{d} = R_H \frac{I_S B}{d} \qquad (6-3)$$

So Hall voltage U_H is proportional to the product of I_S and B, and is inversely proportional to the thickness of Hall material d. Proportional coefficient $R_H = 1/nq$ is called Hall coefficient, which is an important parameter to reflect the capability of material to generate Hall effect.

Hall device is the electric-magnetic transition element made by using the above Hall effect. As for finished Hall device, R_H and d are known; therefore, in actual application, formula (6 – 3) is rewritten into

$$U_H = K_H I_S B \qquad (6-4)$$

Where $K_H = R_H/d$, called the sensitivity of Hall device, which indicates the Hall voltage output of the device under unit operating current and unit operating magnetic strength. In formula (6-4), the unit of operating current I_S is mA, the unit of magnetic strength B is kGs (1Gs = 10^{-4} T), the unit of Hall voltage U_H is mV, and the unit of Hall device sensitivity K_H is mV/(mA·kGs). In accordance to formula (6-4), K_H is known and I_S is adjustable. Magnetic strength B can be found out when U_H is measured, i.e.

$$B = \frac{U_H}{K_H I_S} \tag{6-5}$$

(2) The measurement of Hall voltage U_H

When Hall effect is generated, many secondary effects appear simultaneously. The voltage between 3 and 4 is not strictly equal to Hall voltage U_H, and includes additional voltage resulted from many secondary effects; therefore, we must try to eliminate them. Based on the mechanism generated by secondary effects (see the appendix of the experiment), with the symmetric measurement method of current and magnetic reversal, the influence of secondary effects can basically be eliminated. The practical method is: Keep the values of I_S and B, change the directions of current I_S and magnetic field B, measure the voltage values of four groups U_1, U_2, U_3 and U_4 in turn, i.e.

$+I_S$	$+B$	U_1
$+I_S$	$-B$	U_2
$-I_S$	$-B$	U_3
$-I_S$	$+B$	U_4

and then find out the algebraic average value of the above four groups voltages of U_1, U_2, U_3 and U_4, i.e.

$$U_H = \frac{1}{4}(U_1 - U_2 + U_3 - U_4) \tag{6-6}$$

(3) The magnetic strength in long and straight carrier solenoid

Close winded solenoid can be approximately considered as a series of juxtaposition combination of circle coils with coaxial wires. Therefore, the magnetic strength on a point of a long and straight solenoid axis can be obtained by integration summation of the magnetic strength produced at the points on axis for circle currents. For limited long solenoid, the magnetic strength is maximum at the centre point with equivalent distance to both ends, i.e.

$$B_0 = \frac{\mu_0 N I_M}{\sqrt{L^2 + D^2}} \tag{6-7}$$

Where μ_0 is permeability of vacuum, N is total turn number of solenoid coil, I_M is the exciting current passing through the coil, L is solenoid length, and D is the average diameter of solenoid coil.

In this experiment, $L = 0.28$ m, $D = 0.04$ m, so $\sqrt{L^2 + D^2} \approx L$, and formula (6-7) can be simplified into

$$B_0 = \mu_0 n I_M \qquad (6-8)$$

Where $n = \dfrac{N}{L}$ is the coil turn number of solenoid unit length.

It can be known from the distribution of the lines of magnetic force for long and straight solenoid as shown in Fig. 6-4 that the lines of magnetic force in cavity is basically parallel to the axis. Suppose magnetic strength on axis is B_0; when gradually entering into the port, these straight lines are discrete to both sides. This shows that the magnetic field in cavity is basically uniform, and only on both ends and outside, non-uniformity appears obviously. Based on theory calculation, magnetic strength on one end of long and straight solenoid is 1/2 of magnetic strength on the axis in cavity, i.e. $B_0/2$.

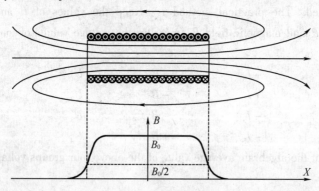

Fig. 6-4 Distribution diagram for magnetic strength of long and straight solenoid

3. Experiment purposes

Grasp the mechanism of Hall effect and the principle of measuring magnetic field; learn how to measure Hall voltage and the method to eliminate secondary effects; use Hall effect method to survey axial magnetic strength distribution of long and straight solenoid, and know the application of Hall effect.

Measurement of Solenoidal Magnetic Field by Hall Effect Method Experiment 6

4. Experiment instruments

The instruments used in this experiment include solenoid magnetic field experiment unit, and solenoid magnetic field tester, as shown in Fig. 6 – 5.

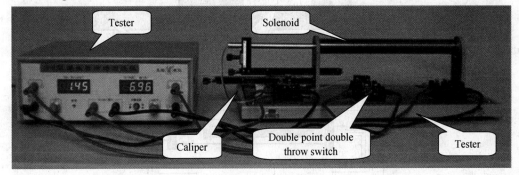

Fig. 6 – 5 Experiment instruments

(1) Solenoid magnetic field experiment unit

Fig. 6 – 6 is the front schematic diagram of solenoid magnetic field experiment unit. The position of Hall device in the solenoid is adjusted by two measuring scales (see Fig. 6 – 7). Note: The position for longitudinal adjusting support Y has already been adjusted, and it is not necessary to adjust it during the experiment; otherwise Hall device will offset from the axis. Fig. 6 – 8 is the wiring diagram.

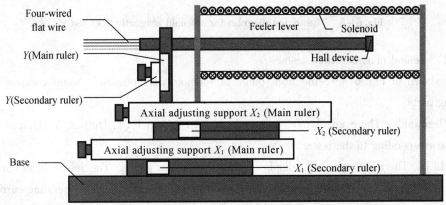

Fig. 6 – 6 Schematic diagram for solenoid magnetic field experiment unit

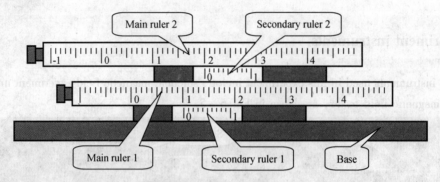

Fig. 6 – 7 Measuring scales

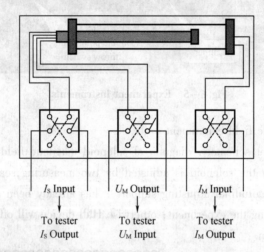

Fig. 6 – 8 The wiring diagram for solenoid magnetic field tester

(2) Solenoid magnetic field tester

As shown in Fig. 6 – 9, the front panel of solenoid magnetic field tester composes of the following parts:

①Terminals. There are 3 pairs of terminals on the panel (I_S Output, I_M Output and U_H Input) corresponding to the tester.

②LED. There are two LED display indicators on the panel. The left indicates measured value (i.e. U_1, U_2, U_3 and U_4); the right is used to display the values of operating current I_S or exciting current I_M adjusted by adjusting knob ("I_S Adjust" or "I_M Adjust").

③Adjusting knob and adjusting screw. On the panel there are two adjusting knobs and an adjusting screw. In the experiment, adjusting knob "I_S Adjust" and "I_M Adjust" separately change the values of operating current I_S and exciting current I_M, and the adjusting screw is used to set zero of LED display indicator before the experiment.

④"Measure selection" button. When the button springs up, the right LED displays the value of operating current I_S; when the button is pressed down, the right LED displays the value of exciting current I_M.

Note: During the operation, observe the state of "MEASURE SELECT" button to avoid adjusting the adjusting knob incorrectly or having wrong reading of data.

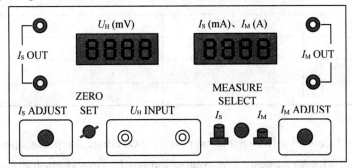

Fig. 6 – 9 The panel diagram for solenoid magnetic field tester

5. Experiment content and operation key points

(1) The measurement of the output characteristics of Hall device

①Connect 3 groups of wires between tester and the experiment unit according to Fig. 6 – 8. Note: Never connect the exciting current "I_M Output" of the tester to "I_S Input" or "U_H Output" of the experiment unit; otherwise, Hall device will be damaged if the power turns on! Therefore, after connecting wires, turn on the power of the tester after being examined by your teacher.

②Turn the knobs of the distance ruler X_1, X_2 on Hall device, slowly move Hall device to the centre position of solenoid ($X_1 = 14.00$ cm, $X_2 = 0.00$ cm).

③Plot $U_H - I_S$ curves. During the process of measuring, keep the exciting current I_M unchanged ($I_M = 0.800$ A), change the values of operating current I_S, use symmetric method to measure the corresponding U_1, U_2, U_3 and U_4, record them in Table 6 – 1, and use formula (6 – 6) to calculate the corresponding values of U_H.

Table 6-1   Recording table for surveying $U_H - I_S$ curve data ($I_M = 0.800$ A)

I_S/mA	U_1/mV $+I_S, +B$	U_2/mV $+I_S, -B$	U_3/mV $-I_S, -B$	U_4/mV $-I_S, +B$	$U_H = \frac{1}{4}(U_1 - U_2 + U_3 - U_4)$
1.00					
2.00					
3.00					
4.00					
5.00					
6.00					
7.00					
8.00					

④Plot $U_H - I_M$ curves. During the process of measuring, keep the operating current I_S unchanged ($I_S = 8.00$ mA), change the exciting current I_M, use symmetric method to measure the corresponding U_1, U_2, U_3 and U_4, record them in Table 6-2, and use formula (6-6) calculate the corresponding values of U_H.

Table 6-2   Recording table for surveying $U_H - I_M$ curve data ($I_S = 8.00$ mA)

I_S/mA	U_1/mV $+I_S, +B$	U_2/mV $+I_S, -B$	U_3/mV $-I_S, -B$	U_4/mV $-I_S, +B$	$U_H = \frac{1}{4}(U_1 - U_2 + U_3 - U_4)$
0.100					
0.200					
0.300					
0.400					
0.500					
0.600					
0.700					
0.800					

(2) The measurement of the distribution of magnetic strength on solenoid axis

Take the operating current $I_S = 8.00$ mA, and the exciting current $I_M = 0.800$ A. During the

Measurement of Solenoidal Magnetic Field by Hall Effect Method Experiment 6

process of measuring, keep the above two physical quantity values unchanged, change the values of X_1 and X_2, use symmetric method to measure the corresponding U_1, U_2, U_3 and U_4, record them in Table 6-3, and use formulas (6-5) and (6-6) to calculate the corresponding Hall voltage U_H and magnetic strength B.

Table 6-3 Recording table for measuring magnetic strength distribution data on solenoid axis ($I_M = 0.800$ A, $I_S = 8.00$ mA)

X_1/cm	X_2/cm	X/cm	U_1/mV $+I_S, +B$	U_2/mV $+I_S, -B$	U_3/mV $-I_S, -B$	U_4/mV $-I_S, +B$	U_H/mV	B/kGs
0.00	0.00							
0.50	0.00							
1.00	0.00							
1.50	0.00							
2.00	0.00							
5.00	0.00							
8.00	0.00							
11.00	0.00							
14.00	0.00							
14.00	3.00							
14.00	6.00							
14.00	9.00							
14.00	12.00							
14.00	12.50							
14.00	13.00							
14.00	13.50							
14.00	14.00							

(3) Notes

①Before turning on the tester, turn counterclockwise the knobs of "I_S Adjust" and "I_M Adjust" to the bottom to make their output currents tend to the minimum status.

②After turning on the tester power, preheat the tester for several minutes.

③During the experiment, if the Hall probe on the tester is adjusted from the right end to the left end, first adjust distance measurement ruler X_1 (the lower end). After X_1 is adjusted to 14.00 cm, adjust distance measurement ruler X_2. In contrary, if the probe is adjusted from the left end to the right end, first adjust distance measurement ruler X_2 (the upper end). After X_2 is adjusted to 0.00 cm, adjust distance measurement ruler X_1. Additionally, the adjustment range of X_1 and X_2 is 0.00 ~ 14.00 cm, and it is not allowable to exceed this range during the operation to avoid the damage of Hall device. Note: Never roughly operate the devices to avoid the damage of instruments.

④When using symmetric method to measure, take care of the direction of 3 double-point double-throw switches. The double-point double-throw switch of "U_H Output" (middle) must always be kept in forward direction (forward), while the double-point double-throw switches of "I_S Input" (left side) and "I_M Input" (right side) must change direction continuously in the experiment.

⑤Tables 6-1, 6-2 and 6-3 must use lateral measurement sequence, i.e. keep the current and position unchanged, in turn measure 4 measurement values U_1, U_2, U_3 and U_4, and then measure continuously the next line.

⑥When using double-point double-throw switches, press them tightly to avoid disconnection or touching resistor.

⑦Before turning off the tester, turn counterclockwise the knobs of "I_S Adjust" and "I_M Adjust" to the minimum.

6. Data recording and processing

(1) Drawing up $U_H - I_S$ curve of Hall device

Use coordinate paper to trace point or Excel to draw up the curve.

(2) Drawing up $U_H - I_M$ curve of Hall device, and using graphic method to find out K_H

①Use coordinate paper to trace point or Excel to draw up the curve.

②Use graphic method to find out K_H.

Substitute formula (6-8) into formula (6-4), and obtain

$$U_H = K_H I_S \mu_0 n I_M \tag{6-9}$$

Which shows that U_H and I_M are linear relation; therefore

$$K_H = \frac{1}{\mu_0 n I_S} \cdot \frac{\Delta U_H}{\Delta I_M} \tag{6-10}$$

Since μ_0 and I_S are known and n is given on the instrument, use graphic method to find out straight line slope $\Delta U_H / \Delta I_M$, and substitute it into the above formula to find out K_H.

(3) Drawing up magnetic strength distribution curve on solenoid axis

①Use a coordinate paper to trace point or Excel to draw up the curve. Take center position O with the equal distance to the ports of two ends of solenoid as coordinate origin, so the distance of Hall device to the center position O point is

$$X = 14.00 - X_1 - X_2 \quad (6-11)$$

②Compare measured magnetic strength B of solenoid center with theoretic value B_0, and find out relative error

$$E_r = \frac{|B - B_0|}{B_0} \times 100\% \quad (6-12)$$

③Verify the magnetic strength of solenoid port is as half as that of the center.

7. Analysis and questions

(1) How does it affect the measured result, if the direction of measuring magnetic strength B is not on the normal direction of conductor board?

(2) Exemplify to describe a method to obtain uniform magnetic field.

(3) How to measure geomagnetic field by using Hall device? Briefly describe the measurement principle and method.

(4) Use Hall device to design vehicle milemeter, and give a brief design plan.

(5) How to measure the carrier concentration, conductivity, and movability of semiconductor material; distinguish conductive type of semiconductor materials.

8. Appendix

(1) Secondary effect of Hall effect

①Non-equipotential effect

It is because the positions of 3 and 4 poles of Hall device are not on an ideal equipotential surface. As shown in Fig. 6 - 10, even if no magnetic field is added, as long as current I_S passes, there is voltage $U_0 = I_S r$, where r is the resistance between two equipotential surfaces on which 3 and 4 are. As the result, when measuring U_H, U_0 is overlapped so that U_H is higher or lower. Obviously, the symbol of U_H depends upon the directions of both I_S and B, and the symbol of U_0 is only related to I_S; therefore, elimination can be done by changing I_S direction.

②Ettingshausen effect

Current is a macro statistic quantity, so formula (6 – 2) uses average velocity $\bar{v}$, and Lorentz force expresses the actual force received by each charge, and in formula (6 – 1) actual velocity v of each charge is used. Obviously, under actual condition, charge motion speed is not uniform with average velocity, i. e. there are carriers higher and lower than average velocity. As shown in Fig. 6 – 11, Lorentz force received by a charge moving with average velocity just cancels the action force of Hall electric field, and the charge whose speed is higher or lower than average velocity separately deflects to the opposite side. With the kinetic energy of carrier converting into heat energy, thus in Y direction differential temperature $T_3 - T_4$ generates; on both poles 3 and 4, thermoelectric effect generates; additional voltage U_E is introduced, and $U_E \propto I_S B$, so that its symbol is always the same as Hall voltage U_H; therefore, symmetric method can not be used to eliminate it, but the introduced error is very small, which can be ignored.

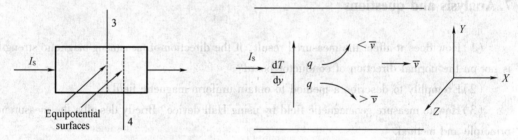

Fig. 6 – 10 Mechanism of non-equipotential voltage generation

Fig. 6 – 11 Principle of ettingshausen effect

③Nernst effect

As shown in Fig. 6 – 12, for the contact resistance of current leads on both ends of device is not equal, different joule heat produces at two contacts after the power is turned on, so that in X direction temperature-gradient appears, which causes carriers to diffuse along the gradient direction, and thus diffusing current generates. By the action of magnetic field of Z direction, heat flow Q which is similar to Hall effect generates an additional electric field δ_N in Y direction, and the symbol of relative voltage U_N is only related to B direction, so it can be eliminated by changing the direction of B.

④Righi-Leduc effect

As shown in Fig. 6 – 13, since carrier speed obeys statistic distribution, heat diffusion current of X direction described in ③, which has the same theory with that described in ②, by the action of magnetic field of Z axis direction, generates temperature gradient in Y direction. The temperature gradient will introduce additional voltage $U_{RL} \propto QB$, and the symbol of U_{RL} is only related to the direction of B, which can be eliminated by changing the direction of B.

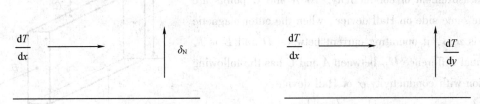

Fig. 6 – 12 Schematic diagram of Nernst effect 　　Fig. 6 – 13 Schematic diagram of Righi-Leduc effect

(2) The elimination method of secondary effect

As stated above, besides U_H, the voltages between both 3 and 4 poles measured in the experiment still include U_0, U_E, U_N and U_{RL}, among which U_0, U_N and U_{RL} can be eliminated through symmetric method (change the direction of I_S and B).

When $+I_S$ and $+B$, $U_1 = U_H + U_0 + U_N + U_{RL} + U_E$
When $+I_S$ and $-B$, $U_2 = -U_H + U_0 - U_N - U_{RL} - U_E$
When $-I_S$ and $-B$, $U_3 = U_H - U_0 - U_N - U_{RL} + U_E$
When $-I_S$ and $+B$, $U_4 = -U_H - U_0 + U_N + U_{RL} - U_E$

From the above 4 formulas, we can obtain

$$U_H + U_E = \frac{1}{4}(U_1 - U_2 + U_3 - U_4)$$

Since the symbol of U_E is always the same as that of U_H, there is no way to eliminate it. But under the condition of not big current and not strong magnetic field, U_E is much less than U_H (only 5% of the latter), so U_E can be ignored, and Hall voltage is

$$U_H = \frac{1}{4}(U_1 - U_2 + U_3 - U_4)$$

(3) Measuring the parameters of semiconductor material by using Hall effect

①Measure the carrier consistency n of semiconductor material

From Hall coefficient definition, we can obtain

$$n = \frac{1}{R_H q}$$

This formula can judge conductive type of semiconductor.

②Measure conductivity σ of semiconductor material

In order to measure conductivity of Hall device, wiring method as Fig. 6 – 14 is generally used for Hall device, in which D and E are current input, A and A' are Hall voltage output, and C is introduced for the measurement of conductivity. As A and C points are on the same side on Hall device, when the outer magnetic field is zero, if operating current between D and E is I, potential difference U_{AC} between A and C has the following relation with conductivity σ of Hall device:

$$\sigma = \frac{1}{\rho} = \frac{I}{U_{AC}} \cdot \frac{l}{S}$$

Where ρ is material resistivity, l is the distance between A and C, and S is sectional area ($S = bd$) of Hall device.

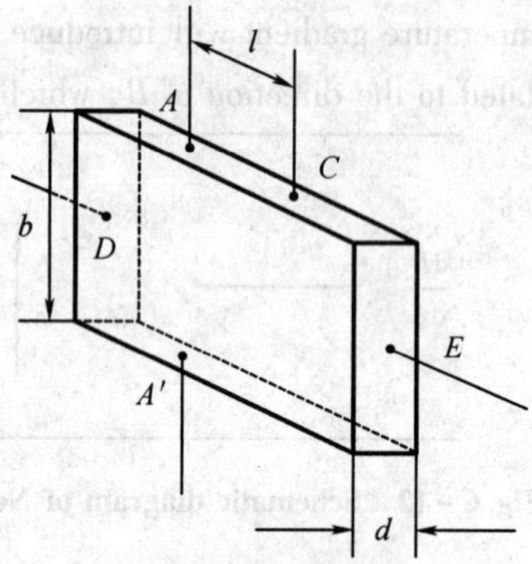

Fig. 6-14 Wiring diagram of measuring conductivity

③Measure movability μ of carriers

Movability reflects the moving ability of carrier in conductive material under the action of outer electric field, and expresses carrier average drift velocity by the action of unit electric field.

Solid physics theory can prove that for semiconductor material, $\sigma = ne\mu$. The following can be obtained

$$\mu = \frac{\sigma}{ne} = |R_H|\sigma$$

(4) The discovery of Hall effect

Hall (Edwin Herbert Hall, 1855—1938) was born in North Gorham, Maine, USA on Nov. 7, 1855, and graduated with excellent result from Bowdoin Institute in 1875. His interest went to scientific research after teaching two years. When talking about his motivation from teaching to science, he said "After having two years teaching career, I went to science to get advanced and also meet the perfect standard in knowledge and moral, but because I have strong enthusiasm for scientific cause, I don't think I have some special talent". Hall described frankly that studying science was due to reality needs, and also making his own knowledge more perfect. He went to postgraduate institute of Johns Hopkins university, and learnt physics with Professor Rowland (Henry Augustus Rowland, 1848—1901). In 1879 he found out "Hall effect", which was his research result of his academic degree thesis, and obtained doctor's degree in 1880. He worked in Hopkins University for one year, and went to Europe for visiting in the summer of 1881, visiting Hermann von Helmholtz's laboratory. During this period, he finished the measurement of Hall effect on some metallic materials. In autumn, he went to Harvard University to be a lecturer; in 1888, an assistant professor; in 1895, a professor; in 1911, he was selected into national academy of science; in 1921, he became honor retired professor. Till the year of 1938, soon before his passing away, he still worked in the laboratory of Harvard University.

Besides the deep research of Hall effect, Hall's main research direction was about heat phenomenon, such as metallic heat conduction, liquid heat action and different thermoelectric effect, especially Thomson effect. During early days of teaching in Harvard University, he prepared the teaching requirements for physics experiment to raise experiment quality of new students, and established 40 physical experiments for high school students. The devices used were simple, the training effects of students were obvious, and the influence was great. Later he wrote some works discussing basic physical education. In 1937, he received "the outstanding contribution medal of Physical teacher" of American physical teacher institute, and became the first honor member of the institute.

Professor Rowland, Hall's supervisor, (Henry Augustus Rowland, 1848 – 1901) (see Fig. 6 – 15) was a brilliant physicist, the first chairman of American Physics Institute, and one of the founders of American Physics Institute. In 1876, he did magnetic effect experiment of charged turn plate in Helmholtz's laboratory and it verified firstly by experiment that motion charge can generate magnetic field. He established the best laboratory of USA in Hopkins University at that time and conducted a series of experiments and researches (measuring heat equivalent value of work, standard value of resistance—ohm, and relation of specific heat of water varying with temperature, etc.). His

Fig. 6 – 15 Rowland

most famous contribution was the development of diffraction grating, providing precision instrument for spectrum measurement and analysis, and promoted the development of spectroscopy. He explained Hall effect by using classics theory. In 1881, Rowland explained magneto-optic rotation phenomenon based on Hall effect, and he thought that the latter was the result of the conduction current in metal rotated by the action of magnetic field, and the former was the result of displacement current in medium rotated in the same condition, so the same rotation equation as Maxwell was derived from mathematics.

Because of his distribution to physics development, he was successively elected as an academician of National Academy of Science of American, a member of the Royal Society, and a foreign academician of Academy of Science of France. Rowland passed away in 1901.

When reading the related part of the book, "Electricity and Magnetic" of Maxwell, Hall took care of a piece of words: "It must remember carefully that mechanic force acted on conductor through which line of magnetic force passed and current flowed, not acted on current but acted on the conductor through which current passed". Hall thought the thesis of Maxwell was contradictory to the direct imagination when people considered this case. The conductor without current will not receive magnetic action, but the acting force received by the conductor through which current passed is proportional to the value of current, and the value of acting force is not related to the dimension and material of metallic wire.

Just soon he read an article "Unipolar Induction" published by Erik – Edlund, a physicist from Sweden, in "Philosophy Magazine (Phi. Mag)" in 1878. The author clearly pointed out that the magnetic field which was acting on the current in a fixed conductor was completely the same as it was acting on the conductor moving freely. He found that the two scientists had different views, and learnt it from his supervisor, Professor Rowland. Rowland said that he doubted the reality of the thesis of Maxwell, and did the experiment for this hurriedly before, but it was not successful.

Hall firstly repeated the experiment of Rowland as shown in Fig. 6 – 16: place a metallic disc between two poles of an electric magnet; make the disc plane perpendicular to the direction of line of magnetic force; and make current flow along a diameter of the disc. Two inputs of a microdetector are connected to the different parts of the disc, through which the value of current is monitored. When electric magnet was not powered, two almost equipotential points were found, at the moment almost no current flew through galvanometer. After exciting current was connected, galvanometer was observed again to detect potential change between the two inputs. In this experiment, any phenomenon was not observed. Later Hall used a narrow tape of gold foil to substitute the metallic disc in accordance to Rowland's suggestion. The experiment was done according to the above thinking, effect of magnetic action was obtained, and the pointer of galvanometer had an obvious deflection, as shown in Fig. 6 – 17. Now it is easy to understand that gold foil is much thinner than metallic disc, and Hall electromotive force is inversely proportional to the thickness of the sample.

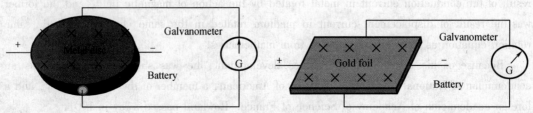

Fig. 6 – 16 Initial experiment device   Fig. 6 – 17 Successful experiment device

The latter research showed that since Hall potential is inversely proportional to the consistency of carriers, the Hall potential in semiconductor materials is much higher than that in metallic materials. Using Hall effect to measure magnetic field is the common used measuring method today. Hall transducers have been widely used in experiment measurement, and Hall effect has become an important method to study electricity performance of semiconductor materials.

(5) Quantum Hall effect

In accordance with classic theory, Hall resistance $R_H = U_H/I_S = K_H B$, i. e. R_H varies continuously with magnetic strength B. But in 1980, Klaus von Klitzing (1943—), a German physicist (see Fig. 6 – 18) observed that at 1.5 K extremely low temperature and in 18.9 T strong magnetic field, when measuring metal-oxide-semiconductor field effect transistor, the Hall resistance varied with the magnetic field, and a series of quantization platform appeared, which was called quantum Hall effect (see Fig. 6 – 19).

Fig. 6 – 18 Klaus von Klitzing

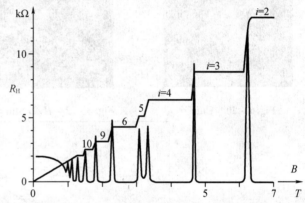

Fig. 6 – 19 Quantum Hall effect

Hall resistance $R_H = U_H/I = h/ne^2$ (h is Planck constant, e is electron quantity of electricity, $n = 1, 2, 3, \cdots$ integer) is not related to the sample and material nature. In 1988, Bureau International des Poids et Measures (BIPM) formally defined the resistance value of the first step ($n = 1$) platform as Klaus von Klitzing constant, the symbol of which is R_k, and stipulated $R_k = 25\ 812.807\ \Omega$ as the standard value of resistance unit.

Quantum Hall effect was the important achievement in condense state physics of 20th century and the development of new technique, for which Klaus von Klitzing won the Nobel Prize for Physics in 1985.

(6) Fraction quantum Hall effect

In 1982, Cui Qi, a Chinese-American physicist (see Fig. 6 – 20), and an German-

American physicist H. L. Stormer (see Fig. 6 – 21) also observed Hall resistance showing quantization platform in the extreme conditions, i.e. the sample with much higher purity and in much stronger magnetic field (20 T), and at much lower temperature (0.1 K), but the extreme difference was that these platforms were not corresponding to the integer but to the fraction, which was called fraction quantum Hall effect (see Fig. 6 – 23). American physicist R. B. Laughlin (see Fig. 6 – 22) explained the discovery through the establishment of model and calculation. For this, the three persons won Nobel Prize for physics of 1998. Cui Qi became the sixth scientist of foreign citizen of Chinese origin, who won the Nobel Prize.

Fig. 6 – 20 Cui Qi   Fig. 6 – 21 H. L Stormer   Fig. 6 – 22 R. B. Laughlin

Fig. 6 – 23 Fraction quantum Hall effect

Experiment 7

Electronic Deflection of Electron Beam and Measurement of the Specific Charge of Electron

1. Background and application

Atom, a basic unit of matter, consists of electron, neutron, and proton. Electron is a kind of elementary particle. Compared with atomic nucleus composed by neutrons and protons, the electron mass is quite small, about 1/1,837 of that of proton. The ratio of the electron charge and the electron mass is called specific charge of electron (e/m), which is one of the basic parameters of charged microparticle. In 1897, Joseph John Thomson (see Fig. 7 – 1) worked in the Cavendish Laboratory of Cambridge University measured the specific charge of electron when studying cathode rays. He studied the deflection of cathode

Fig. 7 – 1 Joseph John Thomson

rays in magnetic field and electric field; consequently he discovered the existence of electrons in the experiment. The discovery of electron and the measurement of the specific charge of electron have been playing an important role in the development of modern physics and are a foundation of studying the substance structure.

Joseph John Thomson (1856—1940) was a British physicist. On Dec. 18, 1856, Thomson was born in Cheetham Hill, Manchester, England. In 1870 he attended Owens College at the young age of 14. At the beginning, he wanted to be an engineer, but made up his mind to study physics influenced by a physics teacher. In 1876, he became a postgraduate of Trinity College,

Cambridge University. Since he was interested in electromagnetic radiation theory of Maxwell, he studied cathode rays and definitely supported corpuscular theory in the debate of the nature of cathode rays. He made a smart experiment to successfully prove the deflection of cathode rays in magnetic and electric fields, which is a crucial evidence to determine that cathode rays are definitely charged particles. Then he confirmed the speed of the particles of cathode rays with the cancellation of electrostatic deflection and the magnetic deflection, measured the mass to charge ratio of these particles and further measured their mass—about 1/1,837 of that of hydrogen atom. Therefore, it can be concluded that the particles of cathode rays are much smaller than atoms; this kind of particles is an elementary particle composing all atoms, and afterwards people named it as electron. Electron is the first kind of elementary particles people recognized. After that, Thomson proposed the atomic structure model—"the atom as being made up of these corpuscles orbiting in a sea of positive charge" (the plum pudding model). Though this model was replaced by the nuclear model of the atom of Ernest Rutherford, it was a beginning of creating the atomic structure model. In 1906, Thomson was awarded a Nobel Prize, "in recognition of the great merits of his theoretical and experimental investigations on the conduction of electricity by gases."

The electromagnetic deflection method adopted by Thomson is to utilize external electric and magnetic fields to make the electron beam deflect. Electromagnetic deflection is used by many modern electronic instruments, for example, the electron beam in kinescope of television deflects controlled by magnetic field. Electromagnetic deflection is also the basic principle of making cyclotron

Fig. 7 – 2   Cyclotron

(see Fig. 7 – 2): charged particles can speed up cyclically until they reach high energy in electric and magnetic fields. At present, the highest energy of the proton obtained by the cyclotron is 30 MeV, and the highest energy of the nitrogen nuclear is 100 MeV.

2. Experiment Principles

(1) Basic structure of oscilloscope tube and the principle of electric focusing

Oscilloscope tube consists of electron gun, deflector and phosphor screen as shown in Fig. 7 – 3. Electron gun is a key part of the oscilloscope tube, and it consists of coaxial metallic cylinders with a filament, a cathode, a grid, an accelerating electrode, first anode, second

Electronic Deflection of Electron Beam and Measurement of the Specific Charge of Electron Experiment 7

anode, etc. When heating current heats the filament and the cathode, the electrons in filament will gain higher kinetic energy and escape from the filament surface. By adjusting the potential of the grid to make it negative potential compared with the cathode, control the amount of electrons passing through it, and thus change the brightness of the spot on the phosphor screen. Accelerating electrode and second anode have the same high voltage, the escaping electrons are accelerated in the electric field to form a beam of electron rays and interact with the fluorescent material on the screen and emits visible light, and thus a bright spot can be seen on the screen. The velocity of electrons in the rays from the "mouth" of the electron gun (the hole of the second anode) is v_z, determined by the following energy equation:

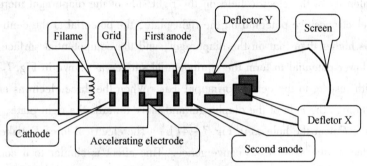

Fig. 7-3 Basic structure of oscilloscope tube

$$\frac{1}{2}mv_z^2 = eU_2 \quad (7-1)$$

Where U_2 is potential difference between the second anode to the cathode, and the final velocity v_z of all electrons emitted from the electron gun is the same, independent of the potential fluctuating of the electrons in the gun.

Accelerating electrode not only can accelerate electrons, but also can make up an electrostatic lens with the first anode and the second anode to focus the scattered electrons.

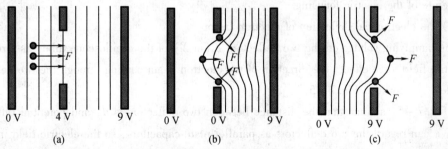

Fig. 7-4 Electric focusing

In order to explain the electric focusing, now put a metallic diaphragm with a circular hole between two parallel charged plate, as shown in Fig. 7-4. In Fig. 7-4(a), put a 4 V voltage on the diaphragm and the electric fields besides the diaphragm are parallel and uniform; the potential of the diaphragm is in the natural potential state, and the electrons emitted by the left plate go through the diaphragm and arrive at the right plate. In this process, the electrons are always in uniformly accelerated motion, with no function of the lens.

In Fig. 7-4(b), since the potential of the diaphragm is zero and lower than "natural" potential, there is no electric field existing on the left of diaphragm far away from the hole; however, the intensity of the electric field on the right side of the diaphragm increases. Because of the effect of electropositive potential on the right plate, the potential at the centre of the hole of the diaphragm is higher than that on the diaphragm, and the equipotential surface extends to the left space with lower potential to form the equipotential surface as shown in Fig. 7-4(b), which is symmetric with respect to the central principal axis. When the three electrons emitted from the left plate pass through the hole of the diaphragm and move towards the right plate, they will suffer an electrical force F near the hole as in Fig. 7-4(b). The electrical force F makes the trajectory of electrons deflect to the axis, i.e. convergence. This effect is similar to a convex lens. The lower the potential of the diaphragm decreases, the more the equipotential surface bends and the more the electric lens converges.

Contrary to Fig. 7-4(b), the potential of the diaphragm in Fig. 7-4(c) is 9 V, higher than the "natural" potential. The equipotential surfaces at the hole of the diaphragm extend to high potential area on the right, and the electrical force F makes the trajectory of electrons deflect away from the axis just like a concave lens, playing a role as a divergence lens.

From the mentioned above, properly design the space distance of the three electrodes to change the potential differences between them, and thus make electron beam converge. This is the principle of the electric focusing.

(2) The electronical deflection of electron beam

When applying voltage on the two deflectors Y (or X) of the oscilloscope tube, as a result of the electric field force effect, the direction of the electron beam passing through the two deflectors will deflect as shown in Fig. 7-5.

In Fig. 7-5, assume that the distance between two deflectors is d, and potential difference is U_y. We can regard the two deflectors as parallel plate capacitors, so the electric field intensity between two deflectors is $E_y = U_y/d$. Because of the effect of electric field force $f_y = eE_y = eU_y/d$, we get the acceleration: $a_y = f_y/m = eU_y/md$. The electrons have no acceleration in the Z-

Electronic Deflection of Electron Beam and Measurement of the Specific Charge of Electron   Experiment 7

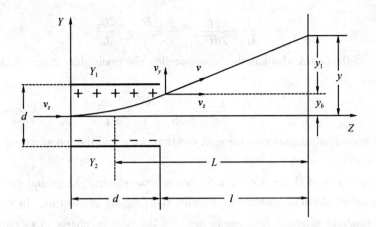

Fig. 7-5 The electronic deflection of electron beam

direction, so the time that the electrons move from the left to the right side of the deflector Y is $t_b = b/v_z$ and the time that electrons move from the right side to the screen is $t_1 = l/v_z$. The vertical displacement when the electrons leave from the right side of the deflector is:

$$y_b = a_y t_b^2/2 = (eU_y/2md) \cdot (b/v_z)^2 \qquad (7-2)$$

and the vertical velocity at the same spot is:

$$v_y = a_y t_b = (eU_y/md) \cdot (b/v_z) \qquad (7-3)$$

When electrons leave from the right side of the deflector, they will not suffer the electric field force, but do a uniform linear motion. The vertical displacement reaching the screen is:

$$y_1 = v_y t_1 = (eU_y/md) \cdot (b/v_z) \cdot (l/v_z) \qquad (7-4)$$

and the total displacement of electrons on the screen is:

$$y = y_b + y_1 = (eU_y/mdv_z^2) \cdot b \cdot (b/2 + l) \qquad (7-5)$$

Make $L = b/2 + l$, i.e. the distance from the centre of the deflector to the screen. Take this formula to the above formula, and according to $\frac{1}{2}mv_z^2 = eU_2$, eventually we have by eliminating v_z:

$$y = \frac{bL}{2dU_2} U_y \qquad (7-6)$$

Formula (7-6) shows that the higher the voltage of the deflector is, the longer the displacement of the spot on the screen will be, and the relationship between them is linear. The proportional constant is numerically equal to the displacement of the spot on screen when the deflection voltage is 1 V. It is called oscilloscope tube's electronic deflection sensitivity S_y, and its reciprocal is called the deflection factor:

$$S_y = \frac{y}{U_y} = \frac{bL}{2dU_2}, \quad \frac{1}{S_y} = \frac{U_y}{y} = \frac{2dU_2}{bL} \tag{7-7}$$

Obviously, deflection X also has the corresponding electronic deflection sensitivity S_x and its deflection factor:

$$S_x = \frac{x}{U_x} = \frac{b'L'}{2dU_2}, \quad \frac{1}{S_x} = \frac{U_x}{x} = \frac{2dU_2}{b'L'} \tag{7-8}$$

(3) The longitudinal magnetic focusing of electron beam and the measurement of the specific charge of electron

The specific charge of electron is the ratio between the electric charge and the electron mass, which is an important physical quantity to describe the property of electron. In this experiment, we use the longitudinal magnetic focusing to measure the specific charge of electron, which is a simple way to measure the specific charge of electron.

Put the oscilloscope tube in a uniform magnetic field of a long straight current-carrying solenoid made by enwinding conductor, and make the direction of the electron beam in the oscilloscope tube and the direction of magnetic flux density B parallel to each other. At the moment, there is no Lorentz force effect on electrons. Electrons do a uniform linear motion with velocity v_z along direction Z, and eventually arrive at the spot O on the screen (as shown in Fig. 7-6). Now, apply a DC voltage U_x between horizontal deflectors X_1 and X_2. After passing through the electric field between the two plates, electrons acquire a transverse velocity v_x perpendicular to B. Because of the Lorentz force, electrons move along a circle in counterclockwise direction when we observe it against the Z direction. The orbital radius of the circular motion is:

$$R = \frac{mv_z}{eB} \tag{7-9}$$

and the period of the circular motion is:

$$T = 2\pi \frac{m}{eB} \tag{7-10}$$

However, when electrons move along the circle, there is a uniform linear motion in the direction of $Z(v_z)$. And after combining these two motions, we find the orbit is spiral. The orbital pitch h is:

$$h = v_z t = 2\pi \frac{mv_z}{eB} \tag{7-11}$$

The different electrons emitted from the first focusing point have the different radial velocities v_r and different circle radii, but if they have same axial velocity v_z, they will move along the different spiral orbits. After moving a distance h, they will refocus at one point. If this point is located on the screen, there will be a spot. And this is the magnetic focusing of electron beam.

Electronic Deflection of Electron Beam and Measurement of the Specific Charge of Electron — Experiment 7

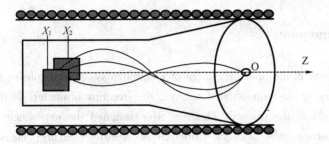

Fig. 7-6 The longitudinal magnetic focusing of electron beam

Set a U_2 (adjust accelerating voltage to change the value of v_z) and select a proper magnetic induction intensity B (adjust the value of the excitation current I of the solenoid), make the orbital pitch h equal to the distance (L) from the first focusing point to the screen; the spot on the screen will become bright, and we have:

$$L = h = \frac{2\pi m v_z}{eB} \tag{7-12}$$

After a transform, we have:

$$\frac{e}{m} = \frac{8\pi^2 U_2}{L^2 B^2} \tag{7-13}$$

The magnetic induction intensity of the solenoid is B. We should calculate by the magnetic field formula of dense multilayered solenoid, but for simplicity, we calculate it by monolayer solenoid formula. The magnetic induction intensity in the middle of the axis is:

$$B = \mu_0 n I \cos\beta$$

$$\cos\beta = \frac{1}{2} \cdot \left[\left(\frac{l}{2}\right)^2 + \left(\frac{D}{2}\right)^2 \right]^{-\frac{1}{2}} \tag{7-14}$$

Where β is the angle between the axis and the radius vector of an arbitrary point on the axis, $\mu_0 = 4\pi \times 10^{-7}$ H·m^{-1}; n is turns of the solenoid per unit of length; I is current; l is the length of the solenoid; D is the diameter of the solenoid.

The relevant parameters of the solenoids used in the experiment (n, l, D) are known in the lab. Substitute them into (7-13), and we have:

$$\frac{e}{m} = \frac{8\pi^2 U_2}{l^2 B^2} = 4.24 \times 10^7 \frac{U_2}{I^2} \tag{7-15}$$

According to the above formula, if we measure the accelerating voltage U_2 and excitation current of the solenoid I, we can figure out the specific charge of electron.

3. Experiment purposes

The measurement of the specific charge of electron plays a vital role in the development of modern physics, and is the foundation of studying the structure of matter. In this experiment, we can learn the thought of the scientist Thomson who designed the experiment; master the basic structure of oscilloscope tube and the principle of electric focusing; measure the electronic deflection sensitivity of the oscilloscope tube; acquaint the basic theory of longitudinal magnetic focusing of electron beam; observe the magnetic focusing; and learn how to measure the specific charge of electron by magnetic focusing.

4. Experiment instruments

Experiment instruments: electric source, solenoid and the specific charge of electron detector as shown in Fig. 7 – 7. The function of each part of the excitation power is show in Fig. 7 – 8. The function of each button on the specific charge of electron detector is shown in Fig. 7 – 9.

Fig. 7 – 7　Experiment instruments

Fig. 7 – 8　Power source

Electronic Deflection of Electron Beam and Measurement of the Specific Charge of Electron

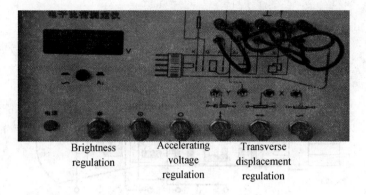

Brightness regulation   Accelerating voltage regulation   Transverse displacement regulation

Fig. 7 – 9 The specific charge of electron detector

5. Experiment content and operation key points

(1) Measure the electronic deflection sensitivity S of the oscilloscope tube
① Connect the wires as shown in Fig. 7 – 10.
② Regulate the oscilloscope tube to observe the electronic deflection.
③ Measure the experiment data and record the data.
④ Drawing on the coordinate paper and figure out S with different accelerating voltages.
⑤ The deflection sensitivity and deflection factor of the oscilloscope tube 8SJ31 are shown as in Table 7 – 1.

Table 7 – 1 The reference values of deflection sensitivity and deflection factor

	deflection sensitivity /(cm/V)	deflection factor /(V/cm)
deflector X	0.035 ~ 0.025	28.6 ~ 40
deflector Y	0.052 ~ 0.038	19.2 ~ 26.3

Note: since the length of the bright line on the screen corresponds to the peak-peak value of the sine voltage, the real voltage can be obtained multiplied the measured voltage by $2\sqrt{2}$.

(2) Measure the specific charge of electron
① Connect the wires as shown in Fig. 7 – 11.

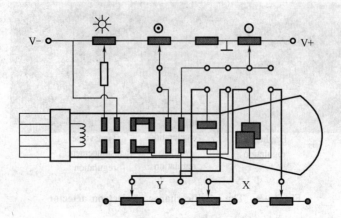

Fig. 7−10 The connection of electronic deflection sensitivity

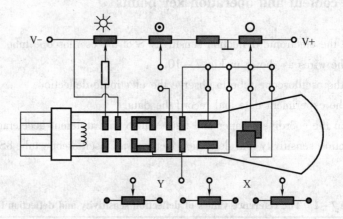

Fig. 7−11 Connection for measuring specific charge of electron

②Select the accelerating voltage U_2.

Note: as we change U_2, the brightness of the spot will change too, so we should regulate the brightness again and avoid making the spot over bright, for this may damage the screen and makes it difficult to judge the performance of focusing. After regulating the brightness of the spot, U_2 will change, and regulate U_2 to the required voltage again.

③Measure excitation current I_1, I_2 and I_3 when it focuses for the first three times. And then convert I_1, I_2, I_3 to the average excitation current I as the first focusing, i.e. weighted mean:

$$I = \frac{I_1 + I_2 + I_3}{1 + 2 + 3}$$

④Reverse the direction of solenoid magnetic field and do the experiment again.

Electronic Deflection of Electron Beam and Measurement of the Specific Charge of Electron　Experiment 7

⑤Calculate the average value of the specific charge of electron, and compare it with the theoretical value $e/m = 1.76 \times 10^{11}$ C·kg^{-1}:

$$\frac{e}{m} = \frac{8\pi^2 V_2}{l^2 B^2} = 4.24 \times 10^7 \frac{U_2}{I^2}$$

6. Data recording and processing

Table 7-2 Data table

Deflection value/cm		1.0	2.0	3.0	4.0	5.0	Deflection sensitivity $S/(\text{cm/V})$	deflection factor (V/cm)
deflector X	900 V							
	1 000 V							
deflector Y	900 V							
	1 000 V							

Table 7-3 Data table

The direction of B	Acceleration voltage U_2/V	Excitation current I/A			Average value	weighted mean	The specific charge of electron e/m ($\times 10^{11}$ C·kg^{-1})
Positive direction	900	I_1					
		I_2					
		I_3					
	1 000	I_1					
		I_2					
		I_3					
Reverse direction	900	I_1					
		I_2					
		I_3					
	1 000	I_1					
		I_2					
		I_3					

The theoretical value $\frac{e}{m} = 1.76 \times 10^{11}$ C·kg^{-1}　$\frac{\overline{e}}{m} =$

$$E_r = \frac{\left|\overline{\dfrac{e}{m}} - \dfrac{e}{m}\right|}{\dfrac{e}{m}} \times 100\% =$$

7. Analysis and questions

(1) How many ways can we use to make the electron beam deflect, what are the principles of these ways?

(2) When observing the changes of the deflection, does the focus of the spot change, why?

(3) Is the numerical value of deflection associated with the brightness of spot, why?

(4) When applying the AC signal on the deflectors, what phenomenon can be observed?

(5) If in the electric chamber the first anode is higher than the second anode, can the focusing be achieved, why?

(6) What influences does the geomagnetic field have on the measuring of the specific charge of electron?

(7) What influences does the space charge effect have on the measuring of the specific charge of electron?

Experiment 8

Determination of Focal Length of Thin Lens

1. Background and application

  Lens is a transparent material (such as glass, crystal, etc.) of an optical element. In astronomy, military, transportation, medicine, art and other fields play an important role.

  The history of lens usage dates back to ancient Greek and Roman times, it has been known at that time and described the lens amplification performance, and found the lens can be concentrated fire. In ancient Rome, people also used lens to help nearsightedness patients to vision correction.

  With the deep study of optical technology, lens has been involved in military defense, aerospace, industrial agriculture, energy, environmental protection, biological medicine, metering, automatic control, and even family life and other fields. Lens, of course, the two most important applications belong to the telescope and microscope. Telescopes make human to realize the "clairvoyance", microscope enables human go into the wonders of the microscopic world (see Fig. 8 – 1). And the camera, and all kinds of optical instruments are inseparable from the lens. Launched in 1990 the Hubble space telescope lens diameter is 2.4 m, angular resolution is about $0.1''$, at an altitude of 615 km outside of the earth's atmosphere around the earth. It adopts computer image processing technology, the image data back to earth. It can be observed 130 light-years distant deep space, found 50 billion galaxies. But scientists still can't meet expectations. Design and manufacture, of the telescope with 8m-diameter lens are in progress and is used to replace the Hubble space telescope (see Fig. 8 – 2), scientists expect to observe "big bang" the beginning of the universe. For microscope, using very short wavelengths of light to improve its resolution is very beneficial. For the optical microscope, using 400 nm violet light irradiation with microscopic observation object, the minimum resolution distance is about 200 nm, the largest magnification is about 2,000. This is the limit of optical microscope. Electrons have volatility, when accelerating voltage of hundreds of thousands of volts, the wavelength of the electron is only about 10^{-3} nm, so the electron microscope can obtain a high resolution. This provides a powerful tool for studying the structure of molecules and atoms.

Fig. 8-1　Early microscope

Fig. 8-2　Hubble space telescope

Now the development of science and technology changes with each passing day, optical instrument has been in the production department and has been widely used in daily life. Although there are many types of optical instruments, lens is a of basic optical element of various kinds of optical instruments. So in order to understand the structure of the optical instruments and correct method of use, we must grasp the lens imaging, learn to light path analysis and adjusting technology. And the focal length is the basic parameters of the lens characteristics, according to different application requirements we can choose appropriate focal length of lens, this will need to determine the focal length of lens. For the deep study of lens imaging, understand the structure of the various common optical instrument this has the positive significance.

2. Experiment principles

The thin lens imaging regularity

Thin lens is one of the most common basic optical element, it is composed of two refractor, a medium of the lens is middle (optical glass). The thickness of the thin lens refers to the center of the lens and compared to the radius of the lens spherical surface it can be neglected. Lens can be divided into two categories: one kind is convex lens, the convergence of light, the shorter the focal length, the greater convergence ability; Another kind is the concave mirror, divergent of the light, the shorter the focal length, the greater divergent ability.

The thin lens imaging relationship can be described below:

$$\frac{1}{p} + \frac{1}{q} = \frac{1}{f} \quad (8-1)$$

The formula can be satisfied by thin lens and paraxial rays (small angle rays which are near the axis). Type and the meaning of symbols and their rules are as image follows: p as the object distance, in the physical and virtual objects is negative; q as image distance, real image is positive, the virtual image is negative; f as the focal length convex, convex lens is positive, concave lens is negative. For thin lens, p and q in the formula are calculated based on the light center of lens.

The most used is convergent lens in the lens, the imaging regularity is expressed by light path in Fig. 8 - 3. Figure in solid indicates physical or real image, the dashed figure with star marks indicates virtual objects or virtual image; For clarity, we draw the two light path diagram, Fig. 8 - 3(a) is the real situation of convex lens imaging, Fig. 8 - 3(b) is the concave lens imaging.

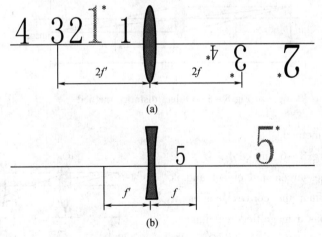

Figure 8 - 3 Imaging rule of lens

The convex lens focal length measurement

(1) The method of plane mirror

If the item will be dispatched to a particular location, the light will be the parallel light, after lens. We if use flat mirror to reflect back the parallel light beam, the light will pass through the lens and image in object surface (see Fig. 8 − 4). At this time the distance between the lens is a lens focal length f.

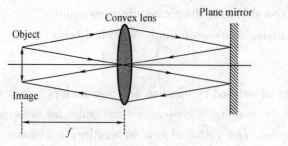

Fig. 8 − 4 Plane mirror method

(2) The object distance method

As shown in Fig. 8 − 5, the light from object A passes, through the convex lens and images on the other side. Using a screen receives this image, measure the distance of p and q. With the formula (8 − 1), We can calculate the focal length f.

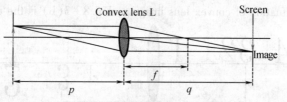

Fig. 8 − 5 Object distance method

(3) Conjugate method

As shown in Fig. 8 − 6, make $A > 4f$, the distance between the screen and object and is kept uncharged. When the convex lens is at O_1, image screen has a magnified real image; Then move the convex lens to the O_2, the screen has a narrow real image. The distance between O_1 and O_2 is d. According to conjugate relations $p_1 = q_2$ and $p_2 = q_1$, we obtain according to formula (8 − 1):

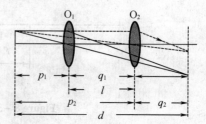

Fig. 8 − 6 Conjugate method

$$f = \frac{d^2 - l^2}{4d} \tag{8-2}$$

The concave lens focal length measurement

(1) The method of plane mirror

As shown in Fig. 8-7, convex lens L_1 creates image D that is a virtual object of concave lens L_2. When adjusting the concave mirror, make *D* in the focal plane, the light after a concave lens is parallel light. With a flat mirror the parallel light beam will be reflected back, then the distance between D and L_2 is the concave lens focal length *f*.

(2) Object distance method

As shown in Fig. 8-8, from A point light through the convex lens converges D after L_1. between L_1 and D we insert concave lens L_2, make L_2D distance less than the focal length of the concave lens L_2 itself, the image of the convex lens is considered to be a concave lens L_2 virtual objects. Based on the concave lens imaging rule, the virtual objects in the concave lens focal length have a real image E. According to the reversibility of light, if put the object in E, then after concave lens, in front of the mirror virtual image D, forms at this time *p* indicates L_2E, *q* means L_2D, in accordance with the rules, the *p* should be positive, the *q* should be negative, depending on the formular (8-1):

$$\frac{1}{p} - \frac{1}{|q|} = \frac{1}{f} \qquad (8-3)$$

So

$$f = \frac{p \cdot |q|}{|q| - p} \qquad (8-4)$$

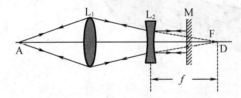

Fig. 8-7   Plane mirror method

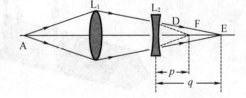

Fig. 8-8   Object distance method

3. Experiment purposes

Understand the definition of a thin lens; master the simple optical path analysis and regulation of technology; deepen the understanding of lens imaging formula; through several methods to quantitatively measure the focal length of the concave and convex lens and master some simple optical path adjustment method; learning about the structure of the optical instruments and correct method of use.

4. Experiment instruments

Thin lens focal length measurement device includes the concave lens, lens, plane mirror, light source, with arrows of screen, image screen, optical bench and tripods.

Optical bench is a multipurpose optical instrument, it is composed of triangular guide and some basic optical components (screen, image screen, bracket, optical bench, etc.). Using this instrument can measure the combination of the focal length of the lens and lens focal length, the regularity of imaging lens. With the appropriate attachments, but also to observe interference, diffraction and polarization phenomenon, as well as to determine the wavelength of light waves and divergence of the laser, microscope, telescope magnification, the luminous intensity of light source, grating constant, etc.

(1) The structure of the optical bench

The structure of the optical bench is shown in Fig. 8 – 9.

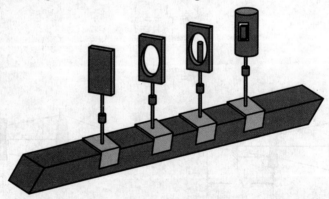

Fig. 8 – 9 Experimental device

(2) "Coaxial" regulation on the optical bench

Optical experiment on the optical bench often before measuring the components should be adjusted to be coaxial. The so-called "coaxial", namely the normal lines of object and image planes coincide with the optical axis of lens, more details should be followed:

①The point of the screen on the lens of the main optical axis, even if the center of things through falls on as the center of the lens.

②The surface and image screen is perpendicular to the main optical axis of the lens, i. e. the axis of the optical bench.

③Optical axis parallel to the optical bench guide rail, to ensure that the components on the guide rail moves without destruction of coaxial condition.

④If there are more than two lenses in the system, first adjust a coaxial system including a lens; then with an additive lens, only adjust the orientation of the second lens, coaxial with the original system, namely the center of the image locates as the center point of the screen. Similarly, put other lenses into the system one by one.

The basic steps of coaxial adjustment can be divided into two steps, coarse and fine adjustment:

(1) Coarse adjustment

The first thing, such as screen, lens in the optical bench, keep them close to each other, with his eyes judgment and adjustment at the center of the various components into roughly in the same line. Screen is vertical to the guide direction, lens optical axis is parallel to the guide rail.

(2) Fine adjustment

Based on law of lens imaging adjustments, if the lens focal length is f, image screen (coarse glass) can be moved to the distance away from the object more than $4f$. So move the lens between the object and image screen, two clear images form (as shown in Fig. 8 – 10). According to the lens movement, adjust the system based on variation of image of the object center on the screen, until the whole system completely satisfy the requirement of "coaxial".

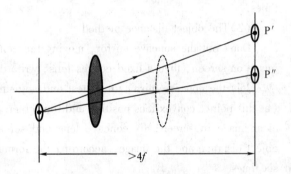

Fig. 8 – 10   Image when $d > 4f$

Notice: Guideway and optical bench must be kept clean, when using, moving and storing care to preserve guideway, avoid heavy pressure on guide rail and prevent deformation of guide rail.

5. Experiment content and operation key points

Convex lens focal length measurement

(1) Flat mirror: At this time, determine screen and lens position coordinates. In order to reduce the accidental error due to judging the clarity of eyes estimated image, repeat the measurement. This experiment repeats measurement 6 times.

(2) The object distance method: To determine the position of screen, lens, image screen coordinates. Requires repeated measurements six times.

(3) The conjugate method: Record the object and screen locations, but must ensure that $d > 4f$. Then when moving lens, corresponding to clear larger and smaller images, record the lens coordinates. Requires repeated measurements six times.

The concave lens focal length measurement

(1) The method of plane mirror

The distance between L_1 and the screen is greater than the focal length of the L_1, similar to convex lens focal length measurement with the flat mirror method, move L_1 and L_2, make the plane mirror reflect light back and form a reverse image clearly on the screen.

Fix the position of L_1, record the position of L_2, removing the L_2 and flat mirror, move image screen along a guide rail mobile screen, until clearly see the image on the screen. Write down image screen location D, the distance between D and L_2 is the f of concave lens. Repeat six times.

(2) The object distance method

Don't put the concave mirror, moving the convex lens along the guide rail, until see clear image on screen, then a fixed convex lens, write down the location D of the image screen.

Join the concave mirror under test and adjustment of its axis, coaxial to the original system (at this point, convex lens position and the screen are unable to move). Move image screen, see clear image in screen, fix concave lens and screen. Respectively take the location, calculate object distance and the image, according the formula to calculate f of the concave lens. Repeat six times.

6. Data recording and processing

According to the experiment content, design the data record form, calculate the focal lengths for a variety of measurement methods and their uncertainty Δf, and present the result forms of $f = f \pm \Delta$.

7. Analysis and questions

(1) From light path diagram in the flat mirror method, we know that the object distance, image distance and the focal length are equal, if three parameters are put into the lens imaging formula, what would happen, meet the thin lens imaging formula? Please give explanations.

(2) By using the distance method measure lens focal length, often take $p = 2f$, the relative uncertainty of the measurement is minimum. Can you prove this conclusion?

(3) Using conjugate convex lens do focal length measurement, why do we need $d > 4f$? Trying to prove it.

8. The appendix

The telescope and microscope (read)

Two main applications of the lens is the telescope and microscope. Let's look at the structure of the telescope and microscope.

(1) The structure of the telescope

Common telescope can be divided into simple Galileo, Kepler telescope and Newton. Galileo invented telescope occupies an important position in the history of the nature of human knowledge. It consists of a concave lens (eyepiece) and a convex lens (lens). Its advantages are simple structure, can directly form an upright image. But after the invention of Kepler this kind of structure has not adopted by professional telescopes, and more be adopted in the toy grade telescope, so called the opera glasses again.

The Kepler telescope consists of two convex lens. Because there is a real image between them, the reticle plate can be conveniently installed, and a variety of excellent performance, so the current military telescope, small telescopes, such as professional telescopes are using this kind of structure. But this kind of structure, imaging is reverse, so to increase the system in the middle for creating upright image.

(2) The main performance parameters of the telescope

The following is the parameters of a telescope, for example to illustrate the meaning of each parameter.

The telescope correct parameter representation:

Specification: 10 × 40, the telescope magnification rate is 10 times, the telescope objective lens is 40 mm in diameter. The zoomed telescope may be marked as: 10 to 40 × 60, then the magnification is between 10 and 40 times continuously adjustable, diameter is 60 mm.

Viewing Angle: 6° 30′, the telescope can watch the sight view, and the larger magnification the telescope, viewing angle is smaller.

1,000 m event horizon: 114/1,000 is another way to tell the view size of a telescope in the 1,000 m distance, the larger magnification the telescope, the smaller the parameter is.

An exit pupil distance: 12 mm, the best distance between eyes and the eyepiece for the image on the retina from the telescope. Associated with the telescope, and generally high magnification telescope with a small exit pupil distance. An exit pupil diameter can be big enough to ensure telescope users wear a gas mask or glasses.

An exit pupil diameter: 4 mm, an exit pupil diameter = diameter/magnification. The 10 × 40 telescope, for example, an exit pupil diameter is 4 mm, of course, this is the ideal state, probably because some telescope optical path design and cost reason the theoretical value can not be achieved.

Resolution: 4.7″, refers to the telescope's ability to distinguish the smallest details. Generally it is related telescope specifications, materials and other factors.

(3) The structure of the microscope

Microscope is an optical instrument formed by a lens or a combination of several lenses. It is the sign of human beings into the atomic age. It is used to magnify tiny objects that people can see by the naked eyes. Optical microscope and electron microscope are typical. Optical microscope in 1590 was invented by the Dutch (see Fig. 8 – 11). The optical microscope can magnify the object 1,500 times now, tell the minimum limit of 0.2 μm.

Many different kinds of optical microscope, such as dark field microscopy, it is a dark field condenser, so that the light beam is not from the central part, but from around toward the specimens. Fluorescence microscope uses ultraviolet light as light source, then objects emit fluorescence. Electron microscope was first assembled in Berlin in 1931. This microscope uses high-speed electron beam instead of the light. Due to the wavelength of the electron much shorter than light waves, so the electron microscope magnification is up to 800,000 times, the minimum limit of 0.2 nm. In 1963 scanning electron microscope can make people see tiny structures on the surface of the object (see Fig. 8 – 12, Fig. 8 – 13 and Fig. 8 – 14).

Fig. 8 – 11 Microscope in 1590

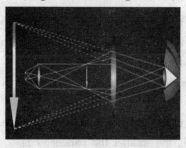

Fig. 8 – 12 The imaging rule of microscope

Fig. 8 – 13 Ordinary optical microscope

The existing various microscopes is composed of the eyepiece and objective, the focal length of the objective lens is very short, the focal length of the eyepiece is very long. The role of the objective is to get magnified real image of object, the role of the eyepiece lens is to take magnified real image as the object, further amplify into a virtual image, its total magnification is the product of the magnification of the eyepiece and the objective lens.

Fig. 8 – 14   Transmission electron microscope

Experiment 9

Adjustment and Application of Spectrometer

1. Background and application

The spectrometer for teaching constantly evolves with the development of science and technology. The structure of spectrometer in textbook is different from foreign physics experiment textbooks in 1930s to "Physical Experiment" written by Westphal and translated by Fushan Wang of Fudan University. In 1930s, the teaching telescope of spectrometer didn't have auto-collimating system, and the spectrometer in the textbook translated by Fushan Wang already have. The spectrometer made in china before 1950s also have auto-collimating telescope (see Fig. 9 – 1). At that time, Gauss eyepiece was installed in auto-collimating telescope. A silk thread was glued on eyepiece as cross-hair relies on manual, it was a basic training of young teachers, and it was the requirements of teaching lessons. After 1960s, both Gauss eyepiece and Abbe eyepiece were installed in auto-collimating telescopes made in china. With the emergence of optical coating technology, reticle began to be made through optical coating technology. After 1980s, optical dial was used in the reading device of spectrometer. So the spectrometer in China is also in constant updates.

Fig. 9 – 1 **Commonly used spectrometers**

Spectrometers usually disintegrate a bunch of multi-wavelength of incident light into monochromatic light; by measuring the angle of the incident light we can get its wavelength information. Because the angle measuring accuracy of spectrometer is higher, it is also sometimes measuring angle precision instruments using optical method. In optical experiments it is often used to determine the direction of the light, and various angles. Some physical parameter as index of refraction, grating constant, dispersive power and so on, often can be determined by directly measuring the angle like minimum deviation, diffraction angle, brewster's angle, so in optical technology, the application of the spectrometer is very extensive. There are many similarities on basic components and regulating principle of spectrometer and other more sophisticated optical instruments, such as the monochromator, spectrograph, etc. So learning and using spectrometer will lay a good foundation for future using more precise optical instruments.

For decades, colleges and universities in our country, the optical experiment teaching, the spectrometer has played an important role. According to our practical teaching experience, spectrometer in experimental teaching has the following four aspects effects.

(1) Operability is strong, there are more than 20 screws can be adjust, it is beneficial to train students' experimental skills. In regulating technology the following training can be made: ①Generation and testing of the parallel light; ②No parallax technical exercises; ③Optical technology practice; ④1/2 regulation technology practice.

(2) Experimental ideas clearly, is conducive to improve the students' experimental skills and level: ①The adjusted collimator is parallel light output, auto-collimator with gauss eyepiece that adjusted output parallel light, and the output of the adjusted and equipped with Abbe eyepiece is not parallel light; ②The experiment content can closely cooperate with the classroom teaching. For example, the dispersion, the minimum deviation angle, the refractive index material and its measurement, diffraction grating and its measurement and the characteristic of light interference, diffraction, etc. ; ③When using a spectrometer to measure the angle, why is there eccentric error and its elimination method; ④How to make the level of the angle under testing parallel to the dial of spectrometer; ⑤When measuring angle, why rotate only one dish between dial and cursor dish; ⑥The physical meaning of 1/2 adjustment method.

(3) Using spectrometer can open multiple teaching experiment: ①The adjustment of spectrometer and prism vertex angle measurement; ②Using the spectrometer to measure the refractive index of glass brick; ③Using the method of minimum deviation angle to measure the refractive index of a prism; ④Tomeasure the refractive index of a prism with grazing incident ray method; ⑤The basic constants measurement of the diffraction grating; ⑥Using spectrometer to observe the phenomenon of the double-slit interference of light waves; ⑦Using spectrometer to observe the phenomenon of the diffraction of light waves; ⑧To observe the phenomenon of waves polarization.

2. Experiment instruments

Spectrometer and supporting the light source, the double-sided mirrors and gratings used in this experiment, as shown in Fig. 9 – 2. The spectrometer is mainly composed of four parts, namely, the telescope, object stage, collimator tube and reading devices (including the dial and reading cursor). In the four components of the spectrometer, in addition to the collimator is fixed to the holder, the other three parts of the instrument can be around the center axis of rotation (also known as the main shaft of the instrument). The method of using each part of the spectrometer is described in the following part.

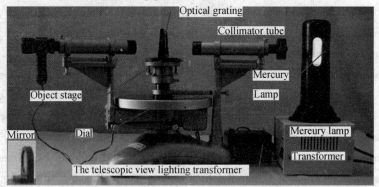

Fig. 9 – 2   Experimental device composed pictures

Telescope

The telescope used by spectrometer is mainly composed of objective lens, eyepiece, partition board and a lens barrel, as shown in Fig. 9 – 3. The telescope tube used on measurement is provided with a reticle. Reticle generally have three horizontal and one vertical lines (usually as a measurement reference). The intersection of the two diameters in circular field is telescope tube axis (the optical axis of the telescope). There is an illuminated green cross in the bottom of the reticle; it can be used to determine the angle between telescope optical axis and the instrument main shaft. This angle can be adjusted with the angle adjusting bolt under the telescope. Users usually need to adjust two focal length, the eyepiece focal length and the focal length of the objective. Because, each person's eye has different focal length. Eyepiece adjustment method is rotating eyepiece sleeve. When you can clearly see the scribed line on the reticle, the focal length of the eyepiece is suitable for your eyes. Adjustment method of the focal length of lens: Loosening the focal length of the objective locking bolt, changing the partition plate and the distance between the lenses.

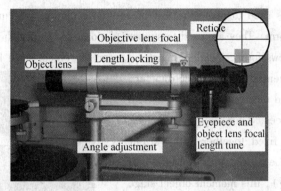

Fig. 9 – 3   The structure and use of the telescope

Collimator tube

A collimator tube is fixed on the spectrometer, as shown in Fig. 9 – 4. It is consist of lens, slit, slit width adjustment, focal adjustment bolt lock and parallel regulation. Lossen focal adjustment bolt lock, stretch out and draw back the slit. When the slit is just right at the focal plane of the lens, the light emitted from collimator is parallel. Angle adjustment can make the optical axis of the collimator perpendicular to the instrument shaft and parallel to the main axis of the telescope. Parallel regulation can adjust the telescope axis and the axis of the collimator in the same line. Dial locking bolts will be illustrated when spectrometer adjusted. Whether the slit width is appropriate and clear will directly affect the accuracy of measurement.

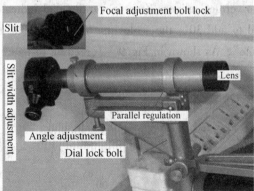

Fig. 9 – 4   The structure and use of the telescope

Object stage

The object stage is used to place the test pieces. The platform is equipped with three lines screws below the platform. They are used to support platform and adjust the angle of platform. The centers of the three screws form a regular triangle. Loosen the object stage lock screw; object stage can rotate around the central axis of spectrometer, it can also move up and down along central axis to adapt different test pieces. Lock object stage lock screw, object will be fixed together with vernier dial. Vernier dial can be locked with dial lock bolt (as shown in Fig. 9 – 5), this moment object stage and vernier dial will be fixed in the same time. Notice, object stage lock screw should be locked all the time except that you need to move object stage up and down.

Fig. 9 – 5 Object stage

Reading device

Reading device is consist of main dial and vernier dial, shown as Fig. 9 – 6. Main dial is divided into 360°. Minimum scale is 0.5°, and you can use vernier reading when scale is less than 0.5°. Review the reading principle of vernier caliper. The minimum degree is minimum scale of main dial divided by the grid number of vernier dial, that is a/n. There are 30 grids in vernier, so 0.5° is divided by 30, and then each grid is 1′. So the reading method of spectrometer reading device is similar to vernier caliper. Zero on vernier dial indicates "degree", and if it is more than 0.5°, you need to add 30′, the scale line on vernier dial aligned with scale line on main dial, indicate "minute". As shown in Fig. 9 – 7, reading in figure a is 81°17′, and reading in figure b is 30°44′.

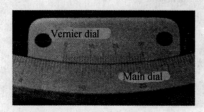

Fig. 9 – 6 Vernier reading

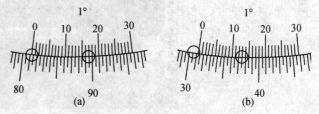

Fig. 9 – 7 Examples of reading

3. Experiment purposes

Spectrometers are used to precisely measure the light space angle, like the space angle between incident light and refracted light, the deflection angle of grating diffraction light and so on. Spectrometers are also used to measure refractive index, angle dispersive power and grating constant. The instrument is more precise, structure is more complex, so you need to adjust it before using it every time. Adjusted steps are more complex and more difficult to master. The main structure and adjusted principle of the spectrometer is similar to monochromator, spectroscope and spectrograph, so noticing its main structure and measurement light path is needed in this experiment, it will lay a foundation for future experiment. So this experiment purpose is: understanding the basic structure and working principle of spectrometer, learning and mastering the adjustment method of the spectrometer, on this basis, grasp the principle and method of calculating wavelength through measuring diffraction angle of grating spectrum, deepen the understanding of grating diffraction theory.

4. The principle and method of adjustment

The so-called spectrometer adjustment, what adjusted? The most important one summed up is: The telescope optical axis is vertical with the main shaft of the instrument, object stage is vertical with the main shaft, collimator tube is vertical with the main shaft, that is three vertical. It also means that, only when the spectrometer meets the three vertical, it can be used to measure various space angle. The simplest method to adjust the spectrometer is adjusting with a double mirror. We will learn this method in this experiment.

Autocollimation method (Principles of Optics)

The optical principle in the process of the spectrometer adjustment is that light travels in straight lines, and angle of incidence is equal to the angle of reflection. The telescope in the spectrometer is also called autocollimating telescope, the autocollimating means that calibrating with light source brought by itself. If the light emitted from the green small cross that has been lighted can image on the symmetry line respect to with the optical axis (That is the intersection of two central lines, o). As shown in Fig. 9 – 8, the light emitted from green small cross and the light reflected can be equivalent to a triangle showed in Fig. 9 – 9. The optical axis is just the high of isosceles triangle. So it is easy to prove that the axis is perpendicular to the reflecting surface by geometric method. Special suggestion: if the light source and reticle scribed line can't be seen clearly, you need to adjust the focal length of the telescope eyepiece; if small cross reflected can't be seen clearly, you need to adjust the telescope lens focal length. Adjustment method refers to telescope in the front. When the telescope focal length adjustment fits your eyes (that is, you can see very clearly), the two focal length do not need to readjust in the process of measuring.

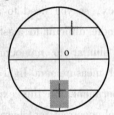 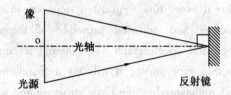

Fig. 9-8 Autocollimation method   Fig. 9-9 Equivalent optical path of autocollimation method

Visual coarse adjustment

To take advantage of autocollimation method to calibrate the telescope optical axis and the plane mirror, first of all, you must be able to see green cross light reflected by the mirror reflection from the telescope's eyepiece. You can control the position of the reflection on the reticle with three platform angle adjustment screws and angle adjustment screw on telescope. Only when the telescope optical axis and the reflection mirror are close to the vertical, reflected image on the reticle can be seen in the eyepiece. Therefore, the so-called coarse adjustment is that making spectrometer basically meet the three vertical conditions with eyes.

The first step of adjustment method is shown in Fig. 9-10, lossen dial lock bolt; second, push the black disk equiped with two dials along the direction indicated by the arrows, then the object stage and the three screws under the platform will rotate around the main shaft (we call this rotating the whole platform); In the rotation process, observe object stage with eyes that whether there is wiggle phenomenon. Adjust platform angle adjustment screws to ensure there is no wiggle phenomenon, and then the object stage is close to be vertical with the main shaft of the instrument. Then put double reflector on the object stage, pay attention to the placement method. As shown in Fig. 9-11, make one side of double mirror face one of the three platform angle adjustment screws. If you do like this, you only need to adjust two platform angle adjustment screws in the rest experiment, and the third one isn't required to be adjusted.

Adjustment and Application of Spectrometer Experiment 9

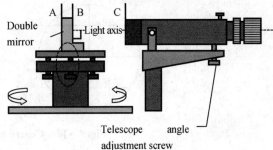

Fig. 9 – 10   Coarse adjustment   Fig. 9 – 11   The method to find reflection

How to find the reflection is the key of spectrometer adjustments. Judgment method of perpendicular between the telescope optical axis and reflective surface is showed in Fig. 9 – 9. The telescope tube is rotated to cross in front of you, and then look from your direction, telescope mirror port can be seen as a straight line C. The telescope optical axis should be basically vertical with C. Put the double mirror on the object stage as shown in Fig. 9 – 11, and A side and B side can also be seen as a straight line; Rotate platform, so that A and B are two reflective surfaces alternately facing the telescope. Observe whether A and C are parallel when A faces with telescope; observe whether B and C are parallel when B faces with telescope; adjust by angle adjustment screw of telescope and platform angle adjustment screws to ensure both B, C and A, C are parallel. Turn the telescope back, observe through eyepiece, rotate platform slightly at the same time to find the reflection of the small cross. If don't find, repeat the above steps. Only when the above conditions are met (the two reflecting surfaces of the double mirror are approximately perpendicular to the optical axis of telescope at this time), you can see the reflected image clearly on the reticle. Because it's only closed to be vertical, the reflected image on the reticle is not necessarily in the positions shown in Fig. 9 – 8. The position shown in Fig. 9 – 12 is not right meeting the condition of autocollimation. We need to adjust in order to move the reflected images of two reflecting surfaces to the position shown in Fig. 9 – 10. Then both A and B are vertical with optial axis of telescope. So the telescope optical axis must be perpendicular to the main shaft of the instrument; if the double side mirror is vertical on the platform, the platform is perpendicular to the main shaft of the instrument (two verticals). We use halved adjustment method to achieve this goal.

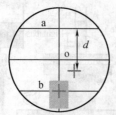 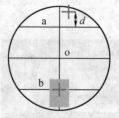

Fig. 9 – 12 Coarse adjustment results

Halved adjustment method

The reticle in the telescope is designed as shown in Fig. 9 – 12. a line and b line are symmetric about the axis o, so as long as the transverse line of reflection is adjusted to the position of a line (as shown in Fig. 9 – 8), the conditions of autocollimation is met. On the basis of coarse adjustment, rotate the object stage to make the telescope aligned respectively the two reflecting surfaces of double mirror. Observe whether the transverse line of reflection is on the position of a line. Generally, the transverse line is above or below a line. At this point careful analysis is needed to determine the adjustment direction. Don't adjust blindly, if a reflection is disappear after adjustment, you must start all over again. Observe the eyepiece of telescope. Assuming the distance between reflected image and a line is d, and then adjust the tilt of the telescope and the stage screws in order to make that the gap is reduced by half. Note: The so-called d reduced by half, is that the stage and the telescope reduce half each tune, that each stage and telescope adjustment is half of the half; rotate the object stage 180°, make the telescope facing the other reflecting surface, repeat the same adjustment method. Repeat adjustments several times until the reflections come from the two reflecting surface are both on the position of a line. This adjustment method is called "halved adjustment method". Note: In the rest experiment, all the parts that can be adjusted in the telescope needn't to be adjusted any more.

Adjustment method of collimator tube

If the optical axis of the collimator tube and the telescope's optical axis can be parallel (or in the same plane), the collimator will be necessarily perpendicular to the optical axis of the main shaft of the instrument (the third vertical); the slit is illuminated by mercury light. If the emergent light of collimator tube coming from slit is parallel, the image of slit on the reticle must be clear and no-parallax.

Adjustment method: ①The axis of the collimators is adjusted to roughly coincide with the optical axis of the telescope from the directions of the side and overlooking. Power on mercury light, fully illuminate slit of the collimator. ② Open slit, observe from telescope and adjust focal length of collimator tube which is the distance between slit and lens until see slit image clearly. Note: if the slit image isn't clear, you need to adjust the focal length of collimator, not the focal length of telescope. Then adjust the width of slit to make its image narrow and bright. ③Rotate the slit to the horizontal position, as shown in Fig. 9 – 13 (a), adjust the gradient of collimator tube to move the slit image to the center line of the telescope reticle, as shown in Fig. 9 – 13 (b). At that time, optical axis of collimator tube and optical axis of telescope must be parallel, and must be perpendicular to the main shaft of the instrument. Then rotate the slit 90°, as shown in Fig. 9 – 13 (c), because we need to measure grating spectrum with slit as light source. So far, all the adjustment is completed and the spectrometer can be used to measure.

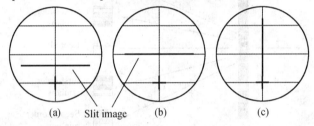

Fig. 9 – 13   Slit adjustment

5. The measuring principle and steps

Grating is a spectral component made according to the principle of multiple slit diffraction. It can make spectrum which is homogeneous and wide spacing. The spectrum made by grating is darker than the spectrum made by prism, but the definition of grating is higher than prism. Grating on the structure can be divided into several kinds like plane grating, stair gratings and concave grating, and at the same time, divided into two categories, transmission type and reflection type.

Transmission type plane grating is choosen in this experiment. It is made through making a large number of parallel to each other, equal width and and spacing indentation on optical glass. When the light shines on grating surface, notch is difficult to get through due to scattering, light can only get though from the slit between the notches. Therefore, the grating is actually a row of dense, uniform and parallel slits.

The principle of measuring wavelength by grating spectrum

Assuming a light with certain wavelength λ exposure perpendicularly to the grating, as shown in Fig. 9 – 14, according the theory of grating diffraction, the position of the bright stripe of diffraction spectrum is determined by the following formula (grating equation):

$$(a+b)\sin\varphi_k = \pm k\lambda$$

or
$$d\sin\varphi_k = \pm k\lambda, k = 0,1,2,\cdots$$

Where $d = a + b$ is called grating constant; λ is the wavelength of incident light; k is bright stripes (spectral lines) series; φ_k is diffraction angle of k series bright stripes.

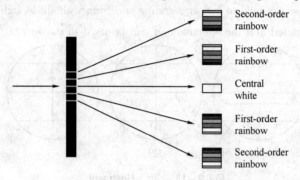

Fig. 9 – 14 Grating diffraction spectrum diagram

If incident light isn't monochromatic light, it can be seen from grating equation that different wavelength light can make different diffraction angle and polychromatic light will be dispersive. In the center, $k = 0$, $\varphi_k = 0$, each monochromatic light overlap together to form a central bright stripes. Symmetrically distributed on both sides of central bright fringe with $k = 1, 2, \cdots$ series spectrum. Spectral lines at all levels are arranged in the order according to the size of the wavelength, and formed a color spectrum, and then polychromatic light is broken down into monochromatic light. And grating spectrum at all levels is symmetrical about the central bright.

If we have known grating constant d and measured diffraction angle φ_1 of a colored bright stripes in spectrum in first series (That is $k = 1$), we can calculate the wavelength λ of the monochromatic light that you have measured with the grating equation.

Measurement method of grating spectrometer diffraction angle

Rotate the whole stage, in order to that we can read dial reading conveniently. Lock the object stage. Put grating on object stage that have been adjusted, the light coming from collimator tube should irradiate vertically the surface of the grating, and slit of collimator and notch of grating should be parallel.

Rotate telescope to observe diffraction spectrum distribution, and ensure that the four spectrum to be measured can be seen clearly from both side of the center bright stripe. In order to eliminate eccentric error of the spectrometer, an angle is needed to be measured by two dial. The reading from the two dial need to be averaged. In order to overcome that the incident light can't ensure normal incidence to the surface of the grating, and it will bring errors to the measurement results, we measure by the symmetry of the grating spectrum. That is we don't measure diffraction angle, but measure the angle between -1 and $+1$ series colored spectrum lines in the mercury lamp spectrum, and then devided by 2 to get the colored light diffraction angle.

6. Data recording and processing

Align vertical line on the reticle to blue, green, yellow 1 and yellow 2 spectrum line in the first left series spectrum respectively, record dial reading both on the left and right, fill in the data record forms. Then align vertical line on the reticle to blue, green, yellow 1 and yellow 2 spectrum line in the first right series spectrum respectively, record dial reading both on the left and right, fill in the data record forms (see Table 9 – 1).

Table 9 – 1　Data record form

	Spectrum / Dial reading	θ_{I}	θ_{II}		Spectrum / Dial reading	θ_{I}	θ_{II}
Blue	$+1$ series			Green	$+1$ series		
	-1 series				-1 series		
Yellow1	Spectrum / Dial reading	θ_{I}	θ_{II}	Yellow2	Spectrum / Dial reading	θ_{I}	θ_{II}
	$+1$ series				$+1$ series		
	-1 series				-1 series		

Fill the following form and the corresponding calculation results item by item (see Table 9 – 2).

Table 9 – 2　Grating constant $d = (a+b) = 1/300$ mm

Data processing \ Spectrum lines	Blue	Green	Yellow1	Yellow2
$\varphi_{\mathrm{I}} = \frac{1}{2}(\theta_{-\mathrm{I}} - \theta_{+\mathrm{I}})$				
$\varphi_{\mathrm{II}} = \frac{1}{2}(\theta_{-\mathrm{II}} - \theta_{+\mathrm{II}})$				
$\bar{\varphi} = \frac{1}{2}(\varphi_{\mathrm{I}} + \varphi_{\mathrm{II}})$				
$\lambda = d\sin\bar{\varphi}(10^{-6}$ m$)$				
$\lambda_{标}(10^{-6}$ m$)$	0.435 8	0.546 1	0.577 0	0.579 1
$\Delta\lambda = \|\lambda_{标} - \lambda\|$				
$E_r = \frac{\Delta\lambda}{\lambda_{标}} \times 100\%$				

7. Analysis and questions

(1) In the experiment if the grating surface is reversed down, or are not placed in the middle of the object stage, will it produce additional error?

(2) How to do to measure grating constant with this experiment device?

(3) What phenomenon will appear when the slit is too wide or too narrow?

Experiment of Equal-thickness Interference

1. Background and application

The interference of light is one of the important optical phenomena, which provides important experimental evidence for the wave property of light. In order to obtain optical interference phenomenon, the two beams must satisfy coherent conditions, i. e. same frequency, same vibration direction and constant phase difference. Since ordinary light sources are not coherent, stable interference field cannot be obtained from two point light sources or two independent parts of a plane light source. In order to guarantee coherent conditions, the method that is usually used is to divide one light beam into two by utilizing a set of optical components and make them meet again after passing different paths. Since the two beams are derived from one beam, they satisfy coherent conditions, and thus a stable visible interference field is formed. There are two ways to decompose the beam:

(1) Wavefront-splitting method

In this method, the wave front of a point light source is divided into two parts and they individually pass through two optical components, overlap after diffraction, reflection or refraction and form an interference field in a certain region. Young's experiment is a model of the wavefront-splitting instruments; many other instruments such as the biprism, Fresnel double-mirror and Lloyd's mirror, etc. are also wavefront-splitting instruments.

(2) Division-of-amplitude method

When a light beam irradiates the interface of two transparent media, a portion of light reflects and the other transmits, so this method is called division-of-amplitude method. The simplest division-of-amplitude interference device is thin film, and other devices include Michelson interferometer, Newton ring (see Fig. 10 – 1), wedge film, etc..

The equal thickness interference of light is an interference phenomenon based on division-of-amplitude, and Newton ring and wedge film are typical devices of the equal thickness interference. In modern precise measurement technologies, equal thickness interference has many applications: it is

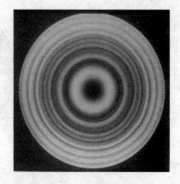

Fig. 10 – 1　Newton ring formed by white light

an important means to test the surface finish and flatness of high-precision optical surface (as shown in the following Fig. 10 – 2); it can also precisely measure the thickness of a thin film and a small angle, measure curvature radius of curved surface and the sample's expansion coefficient, study components' internal stress distribution, etc.

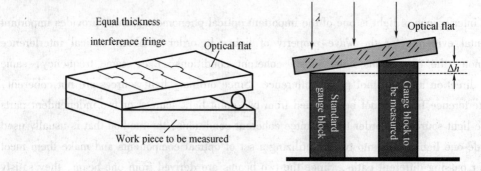

Fig. 10 – 2　Flatness measurement and gauge block calibrating device

2. Experiment principles

(1) Equal thickness interference

When a wedge plate is irradiated by a broad light source, a beam of incident light from the center point of the light source S_0 is divided into two beams after reflected by the two planes of the plate, and the two light beams intersect at a certain point P (as shown in Fig. 10 – 3). The interference effect at point P is determined by the optical path difference of the two light beams, which is expressed as follows:

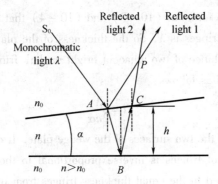

Fig. 10 – 3 The interference at a certain point P in the localized plane of the wedge plate

$$\Delta = n(AB + BC) - n_0(AP - CP)$$

Where n is the refractive index of the wedge plate and n_0 is the refractive index of the surrounding medium. If the thickness of the plate is small, the angle of the wedge is small and the light almost irradiates vertically, the optical path difference can be expressed as follows:

$$\Delta = 2nh \qquad (10-1)$$

where h is the thickness at point B of the wedge plate. Considering the additional optical path difference produced by half wave loss of reflection by the top or bottom surfaces, formula (10 – 1) can be rewritten as follows:

$$\Delta = 2nh + \frac{\lambda}{2} \qquad (10-2)$$

If the refractive index of the wedge plate is uniform, it can be obtained from formula (10 – 2) that the optical path difference of the two reflected beams at the intersecting point only depends on the thickness h at the reflecting point of the plate. Because the interference fringe is a trace of points with the same thickness at the plate, this kind of fringe is called equal thickness fringe and the interference is called equal thickness interference.

If the optical path difference Δ satisfies the following condition:

$$\Delta = 2nh + \frac{\lambda}{2} = m\lambda \quad m = 1,2,\cdots \qquad (10-3)$$

At the point P the light intensity is maximum, and the trace of the points of the same order is called a bright fringe. If the optical path difference Δ satisfies the following condition:

$$\Delta = 2nh + \frac{\lambda}{2} = (2m+1)\frac{\lambda}{2} \quad m = 0,1,2,\cdots \qquad (10-4)$$

At the point P the light intensity is minimum, and the trace of the points of the same order is called a dark fringe.

It can be obtained from formulas (10-3) and (10-4) that the optical path difference of two adjacent bright or dark fringes is λ. So the thickness of the plate changes by λ/2n from one fringe to another and the distance of two adjacent bright or dark fringes (also called the distance of fringes) can be expressed as follows:

$$l = \frac{\lambda}{2n\alpha} \tag{10-5}$$

Where α is the angle of the two surfaces of the wedge plate. It can be obtained from formulas (10-5) that the distance of fringes is inverse proportional to the angle of the wedge α. The conclusion can also be applied to the equal thickness fringes from plates with other shapes.

(2) Measure the curvature radius of lens by Newton ring

Newton ring experiment is an example of the applications of the equal thickness interference. When place a plano-convex lens with a great curvature radius R on a piece of planar glass, a thin air film with a variable thickness increasing from zero will be formed between the convex surface of the lens and the planar glass. When the device is vertically irradiated by a monochromatic light, a group of circular interference fringes centered at the contact point O will be formed, sparse at the center and dense at the outer, which is called Newton ring. The curvature radius of the lens will be calculated by the radius of Newton ring, which can be measured by the reading microscope.

As shown in Fig. 10-4, assuming the central dark spot is zero order and the radius of the m order ring is r_m, we have

$$r_m^2 = R^2 - (R-h)^2 = 2Rh - h^2$$

Where R is the curvature radius of the convex surface of the plano-convex lens and h is the thickness of the air film that corresponds to the m order of ring. Since R is much larger than h, the term h^2 can be neglected, and we have

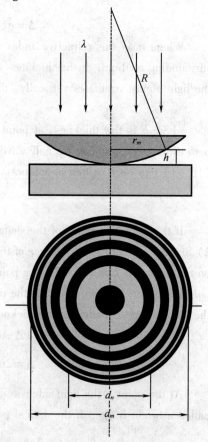

Fig. 10-4 Newton ring interference

$$h = \frac{r_m^2}{2R} \qquad (10-6)$$

Substitute the former formula into the m order ring's optical path difference formula:

$$2h + \frac{\lambda}{2} = (2m+1)\frac{\lambda}{2} \qquad (10-7)$$

Thus

$$R = \frac{r_m^2}{m\lambda} \qquad (10-8)$$

So if the wavelength λ of the monochromatic light is known and the radius of the m order ring r_m can be measured by the reading microscope, the curvature radius of the lens will be calculated by the formula (10-8).

At the center of Newton ring, which is also the contact point of the convex surface of the lens and the planar glass, because h is equal to zero and the optical path difference Δ of two reflected beams is equal to $\lambda/2$, the center of Newton ring is a dark spot. It is obvious that a group of circular interference fringes can also be seen in the direction of transmitted light. The intensity distribution of these fringes is opposite to that of the reflected light, so the center of circular interference fringes of the transmitted light is a bright spot.

In the former discussion, the deformation near the contact point of the plano-convex lens and the planar glass is neglected, and if the deformation is taken into account, the radius of rings near the contact point does not accord with the formula (10-8). Under the circumstances, the curvature radius of the lens can be calculated by measuring the radii of two rings far away from the center. Assume the radius of the m order of ring is r_m and the radius of the n order of ring is r_n, from the formula (10-8), we can get

$$r_m^2 = mR\lambda, \quad r_n^2 = nR\lambda$$

Thus

$$R = \frac{r_m^2 - r_n^2}{(m-n)\lambda} = \frac{d_m^2 - d_n^2}{4(m-n)\lambda} \qquad (10-9)$$

Where d_m and d_n are the diameters of m and n order of rings, respectively. It can be seen from the formula (10-9) that we only count the order difference $m-n$ of the rings regardless of determining the order of the rings. It's easy to prove that the quadratic difference of diameters is equal to that of strings. So it is not necessary to determine the center of the rings and thus we avoid the difficulty of determination of the order and the center of rings.

(3) Measure the thickness of sheet by the method of wedge interference

As shown in Fig. 10-5, when one end of two optical glass plates are placed together and the other end are plugged by a thin wire (a sheet), an air wedge will form between the two glass plates. When the device is vertically irradiated by monochromatic light, the light reflected by the two surfaces of the air wedge will interfere. Assume the thickness of a point E on the top surface of the air wedge is e, and the optical path difference can be expressed as follows:

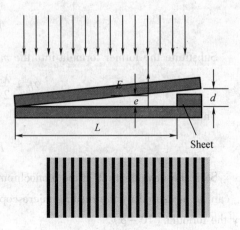

$$\Delta = 2e + \frac{\lambda}{2}$$

When the optical path difference Δ is equal to $(2k+1)\lambda/2$ ($k = 0,1,2,\cdots$), the two beams of

Fig. 10 – 5   Wedge interference

light destructively interference and dark fringes are obtained, so we can get

$$e = k\frac{\lambda}{2} \qquad (10-10)$$

It can be seen from the formula (10 – 10), when k is equal to zero, e is equal to zero and the zero order of the dark fringe will be obtained along the interface of two glass plates. When k is not equal to zero, because the regions with the same thickness on the air wedge are a group of parallel lines that are parallel to the interface of two glass plates, the interference fringes are a group of parallel fringes with the same distance and parallel to the interface. If the $k = N$ order of the dark fringe appears somewhere on the sheet, the depth d of the sheet can be expressed as follows:

$$d = N\frac{\lambda}{2} \qquad (10-11)$$

It can be seen from the formula (10 – 11) that the depth d can be directly calculated by counting the number N of the interference fringes from the interface to the sheet, but the method is difficult to realize. Therefore, we firstly measure the number n of the interference fringes within a unit length and secondly the total length from the interface to the sheet (as shown in Fig. 10 – 5), and finally N is calculated by the following formula

$$N = Ln \qquad (10-12)$$

Considering the small distance between two adjacent interference fringes, n is calculated by measuring the distance between 50 interference fringes. Assume the distance between 50 interference fringes is l, then

$$n = \frac{50}{l} \qquad (10-13)$$

Assuming a certain fringe as the zero order of the fringe, its position x_0 can be read from the reading microscope and the position of the fiftieth order of the fringe can also be read as x_{50}, then

$$l = x_{50} - x_0$$

Substitute the formula (10 – 12) and (10 – 12) into (10 – 11), we can get

$$d = \frac{50L}{l} \times \frac{\lambda}{2} \qquad (10 - 14)$$

3. Experiment purposes

(1) Understand the principle of equal thickness interference.
(2) Master the principles and methods of measuring curvature radius of a sphere with Newton ring and measuring sheet thickness with the wedge interference.
(3) Learn to use reading microscope based on understanding its optical principles systematically.

4. Experiment instruments

Reading microscope, Newton ring device, wedge device, and sodium lamp.
(1) The reading method of the reading microscope

Reading microscope is shown in Fig. 10 – 6: the interval value of the main scale is 1 mm, the interval value of the fine adjustment drum is 0.01 mm and can be estimated to read to 0.001 mm.

Fig. 10 – 6 Reading microscope microscope

Now take Fig. 10 – 7 as the example, and the correct reading result is 15.506 mm.

Main scale: 15 mm Vernier: 0.506 mm

Fig. 10 –7 Demonstration of reading microscope

(2) Sodium lamp

Sodium lamp is a kind of gas discharge lamp, the discharge tube of which is filled with sodium and argon. At the very moment of turning on the lamp, argon discharges and emits pink luminescence. After that, sodium is evaporated and discharges and emits yellow luminescence.

In the visible range, the wavelengths of two sodium spectra are 589.59 nm and 589.00 nm, respectively. These two spectra are close to each other, therefore the light of sodium can be regarded as a monochromatic light source and its wavelength is the average value 589.30 nm.

5. Experiment content and operation key points

(1) Measure the curvature radius of lens with Newton ring

The experiment device is as shown in Fig. 10 –6:

①Before experiment, read the instruction of reading microscope carefully.

②Check Newton ring, and adjust the three screws to make the surface to be measured exactly press the plate plane. Don't over-tighten the screws to avoid damaging the contactile glass components.

③During the measurement, from formula (10 –9), when λ is fixed, the radius of curvature R can be calculated only after the values of d_m and d_n are figured out. Usually $m - n = 25$ for the convenience of calculation.

④As the result of mechanical structure, there will be a backlash error if the reading microscope is used incorrectly. Therefore, during the measurement process, the handwheel can only be rotated in one direction to avoid the backlash error; otherwise, the data are meaningless and invalidated. The correct operating method is: turn the cross-wire leftwards from the central dark spot to the 45th dark circle; then reverse the drum, and start to read the reading when the cross-wire moves back to the 40th dark circle. Write down the positions of the 39th to the 31st dark circles and the 15th to the 6th dark circles on the left in turn; and then write down the positions of the 6th to the 15th dark circles and the 31st to the 40th dark circles on the right in turn.

(2) Measure the sheet thickness with wedge film

①Observe wedge interference phenomenon

Turn on the sodium lamp. Replace Newton ring by wedge device and put it under the objective lens. Similar to the Newton ring experiment, make the monochromatic light normal incidence to the wedges' surface, adjust the focusing knob in the microscope carefully and use the microscope to observe the interference fringe, and then adjust the orientation of the wedge in order to make the cross-wire perpendicular to the fringe. Rotate the fine adjustment drum, and observe the wedge interference phenomenon.

②Measure the sheet thickness

Measure the length L from the wedge interface to the sheet, and L is a single measurement value; when measuring l, select the straight and even distributed part of the fringes, and do the measurement twice to take the average value.

6. Data recording and processing

(1) Measure the curvature radius of lens with Newton ring

①Data record table

The data are recorded in Table 10 – 1:

Table 10 – 1 Data record table

$\lambda = 0.5893 \times 10^{-6}$ m $\Delta_{仪} = 0.005$ mm   单位:mm

Order of ring	m	40	39	38	37	36	35	34	33	32	31
Position of the ring X_m	left										
	right										
diameter	d_m										

Order of ring	n	15	14	13	12	11	10	9	8	7	6
Position of the ring X_m	left										
	right										
diameter	d_m										
$d_m^2 - d_n^2$											

②Data record table

a. process data with the method of the successive minus (see introduction).

b. calculate curvature radius R and its uncertainty degree ΔR according to the measurement data, and the result is

$$R = \overline{R} \pm \Delta R$$

Some formulas of data processing are shown as follows:

$$\Delta d_{m,n} = \sqrt{\Delta_{仪}^2 + \Delta_{仪}^2} = \sqrt{0.005^2 + 0.005^2}$$

$$S_{d_m^2 - d_n^2} = \sqrt{\frac{\sum[(\overline{d_m^2 - d_n^2}) - (d_m^2 - d_n^2)]^2}{n-1}}$$

$$\Delta_B(d_m^2 - d_n^2) = \sqrt{(2d_{40}\Delta d_{40})^2 + (2d_{15}\Delta d_{15})^2}$$

$$\Delta_{d_m^2 - d_n^2} = \sqrt{S^2 + \Delta_B^2}, \Delta R = \frac{\Delta_{d_m^2 - d_n^2}}{4(m-n)\lambda}, \overline{R} = \quad , R = \overline{R} \pm \Delta R$$

$$E_r = \frac{\Delta R}{\overline{R}} \times 100\%$$

(2) Measure the sheet thickness with wedge film

①Make a data record table.

②The demands of data processing:

As known, $L = 23.5$ mm and $\lambda = 0.5893 \times 10^{-6}$ m. Calculate the sheet thickness d according to the formula (10 – 14).

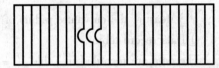

Fig. 10 – 8   Wedge interference

7. Analysis and questions

(1) Why is the distance between the two adjacent dark fringes (or bright fringes) near the center longer than that at the edge?

(2) During the experiment, the center of Newton ring is a bright spot instead of a dark one, and does it have an influence on the experiment result, why?

(3) Use the optical path as shown in Fig. 10 – 5, we can get the wedge interference pattern shown in Fig. 10 – 8. Is the low surface of the glass convex or concave, how to measure it?

Measurement of Sound Velocity

1. Background and application

In the field of scientific research and engineering technology, usually for a variety of measured signals to understand the characteristics and carry by all kinds of information, the oscilloscope is a wide range of electronic measuring instrument, it can directly measure electrical signals, also can through the transducer to convert all kinds of electrical signals into electrical signals to measure. It can display the signal waveform, but also determine the parameters such as amplitude, period and frequency ect. With double trace oscilloscope can also measure time lag or phase difference between two signals. This experiment that it used the oscilloscope measuring sound velocity of oscilloscope is a typical application example.

Sound is a mechanical disturbance in gaseous, liquid, and solid material. Due to its consistent with the direction of propagation direction of vibration, so the sound is a longitudinal wave. Its wide frequency, the range is from 20 to 1~1,010 Hz. Vibration frequency between 20 Hz to 20 kHz sound can be heard by people; more than 20 kHz frequency called ultrasonic sound waves.

Voice in the process of spread can also cause a change of physical optical, electromagnetic, mechanical, chemical properties and the nature of human physiology, psychology and so on, and they will affect the propagation of sound in turn.

Acoustics is the study of sound production, transmission, receive, and process discipline. As a branch of physics is one of the oldest subjects, it has a long history, is a development of the discipline. Vibration is the theoretical basis of the sound source, from Galileo's work to the discovery of hooke's law. In the eighteenth century the development of mathematics, promote the development of the theory of acoustic. The study of mechanical vibration opened up a new direction of the sound vibration research.

At present, the development of the acoustic has widely penetrated into the national economy and national defense and other fields. Sound like water in the field of the development of various types of sonar, ultrasonic imaging in medicine and industry and acoustic weapons such as weapons and equipment of all has close relationship with the development of acoustics. Ongoing research topics include: Ultrasonic motors, electroluminescence, space-time finite wave reflection and transmission on the interface, hall sound reverberation sound waves and sound waves, magnetic problem such as the application in engineering testing. In terms of basic research, was found in the LCD in the study of nonlinear dynamics problem pointing in the direction of wave.

Measurement for acoustic characteristics (such as frequency, wave velocity and wave and sound pressure attenuation and phase, etc.) is an important content of acoustic technology, especially the measurement of acoustic wave velocity (hereinafter referred to as the speed of sound), in sonar, flaw detection and ranging applications has the vital significance.

2. Experiment principles

The structure and working principle of oscilloscope

Oscilloscope specification and model of numerous and internal structure are complex, but the basic structure is similar, all can be roughly divided into the oscilloscope tube (CRT), Y axis amplification (and attenuation) system, X magnification (and attenuation) system, scanning step and the whole system four parts (see Fig. 11 – 1).

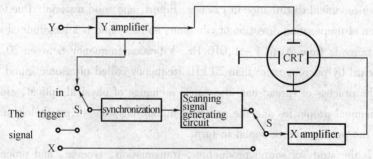

Fig. 11 – 1 Structure of Oscilloscope

The principle of oscilloscope, it briefly as follows: Y amplifier is responsible for enlarge the weak signal under test, trigger synchronous circuit under test to decide when to begin to manifest signal waveform, scanning circuit and signal amplifier X is responsible for the waveform display speed, while the oscilloscope tube (CRT) is a display device. Working process can be qualitatively compared with sand pendulum. Simple pendulum consists of a filled with sand funnel, pendulum swing up, under the uniform sand leakage, after the motion at a constant speed at the bottom of the paper, the sand will be on paper, draw a picture of a harmonic vibration curve habits often referred to as a "wave" curve, as shown in Fig. 11 - 2. Device is placed in the sand, the vibration period of simple pendulum is equivalent to signal under test, the size of a simple pendulum amplitude changes of Y amplifier magnification, paper is equal to the oscilloscope tube (CRT), the speed of paper equivalent to a scanning signal amplifier circuit and X feature set, the starting time on the paper motion control is equivalent to trigger synchronous circuit function. According to the speed of paper (scan) and the length of the "wave", we will be able to measure the period of simple pendulum. The working principle of oscilloscope is much more complex, actually principle readers, please refer to appendix is introduced in detail.

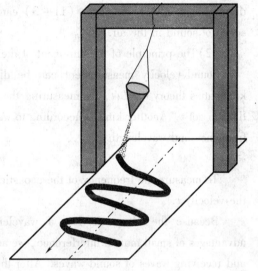

Fig. 11 - 2 Sand pendulum

The principle of sound velocity measurement

(1) Sound waves in the air velocity

In the speed of sound waves in the air (sound) can be written as:

$$v = \frac{1}{\sqrt{\rho X}} \quad (11-1)$$

Where ρ is the density of the gas; X is the compression coefficient. Because of gas compression and sparse part of the spread of fast a; can be considered as adiabatic, so under the ideal gas state approximation, $X = 1/(\gamma p)$, the substitution type (11 - 1):

$$v = \sqrt{\frac{\gamma p}{\rho}} = \sqrt{\frac{\gamma RT}{M}} \quad (11-2)$$

$R = 8.31$ J/(mol · K) as the molar gas constant; M is the molar mass of the gas. $\gamma = c_p/c_V$ for the specific heat ratio; T for thermodynamic temperature. By formula (11 - 2), the sound velocity and the temperature, the molar mass of the gas and the specific heat ratio, after two parameters related to the gas composition. Using the Celsius scale and the conversion relationship between the thermodynamic temperature scale, type (11 -2) can be expressed in Celsius scale:

$$v = v_0 \sqrt{1 + \frac{t}{T_0}} = v_0 \sqrt{1 + \frac{t}{T_0}} \qquad (11-3)$$

Among them, $v_0 = \sqrt{\frac{\gamma R T_0}{M}}$ = 331.45 m/s is the standard state of the velocity of sound in air drying, T_0 = 273.15 K. Type (11 -3) can be used as a theoretical calculation formula of the speed of sound in the air.

(2) The principle of measurement of the velocity of sound in air

Sound velocity measurement can be divided into two kinds: one kind is based on the kinematics theory $v = L/t$, by measuring the acoustic propagation distance L compared with the time to get v. Another kind is according to wave theory, the velocity and the relationship between frequency and wavelength:

$$v = f \cdot \lambda \qquad (11-4)$$

To measure the frequency of the acoustic f and wavelength λ, by formula (11 -4) calculate the velocity v.

Because the ultrasonic has a wavelength, easy to directional transmission, and the advantages of small mutual interference, we adopt piezoelectric ceramic transducer for the emitting and receiving waves of sound waves. After the measured acoustic frequency f and wavelength λ, then v is obtained by formula (11 -4). Type, the frequency of the acoustic f read out directly by the signal generator, the wavelength lambda by the resonant interfering method or phase comparison method were measured.

(3) The resonant interfering method (standing wave) was used to measure the wavelength of sound waves

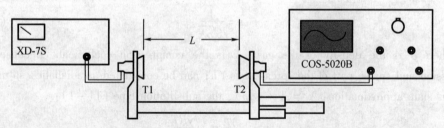

Fig. 11 - 3 The resonant interfering method (standing wave) was used to measure sound wavelengths experiment device

Measurement of Sound Velocity Experiment 11

The resonant interfering method was used to measure sound wavelengths of experiment device as shown in Fig. 11 – 3. The figure of T1 and T2 for piezoelectric ultrasonic transducer. Signal generator output sinusoidal ac signal on T1, completed by T1 electroacoustic transformation, as a source, a wavefront approximation for plane sound wave. T2 as ultrasonic receiving transducer, converting the received acoustic signal into electrical signal, and then access the oscilloscope observation. T2 while receiving sound waves, its surface is also part of the sound wave. When the surface of the T1 and T2 parallel to each other, back and forth between T1 and T2 acoustic interference occurs to form a standing wave.

According to wave theory, set up along the x direction for injection of the incident wave equation:

$$y_1 = A\cos(\omega t - \frac{2\pi}{\lambda}x)$$

For the reflected wave equation:

$$y_2 = A\cos(\omega t + \frac{2\pi}{\lambda}x)$$

Where A is sound amplitude; ω is angular frequency; $2\pi x/\lambda$ for the wave to the coordinates x, the phase change with the time (t). At any time t, in the air of a vibration equation to a location:

$$y = y_1 + y_2 = (2A\cos\frac{2\pi}{\lambda}x)\cos\omega t \qquad (11-5)$$

The type is the standing wave equation.

When $\left|\cos\frac{2\pi}{\lambda}x\right| = 1$, mean $\frac{2\pi}{\lambda}x = k\pi$, at $x = k \cdot \frac{\lambda}{2}$ ($k = 0, 1, 2, \cdots$), Synthesis of vibration amplitude of the largest, known as the loop or acoustic amplitude of great value.

When $\left|\cos\frac{2\pi}{\lambda}x\right| = 0$, mean $\frac{2\pi}{\lambda}x = (2k+1)\frac{\pi}{2}$, at $x = (2k+1) \cdot \frac{\lambda}{4}$ ($k = 0, 1, 2, \cdots$), Synthesis of vibration amplitude of the youngest, referred to as the amplitude of sound wave section or minimum.

These are simple derivation of the theory of the actual situation is much more complicated. According to wave theory, due to the ultrasonic propagation in transducer is much faster than the air in a big way, in the receiver, from sound pressure test is antinode, examination is the nodal from acoustic amplitude. Change the distance between the two transducer, when the distance between them is half wavelength of integer times, two end face of transducer will form a resonant cavity, the sound waves in the cavity between the two end face repeatedly to form stable standing wave back and forth, the transmitting transducer and receiving transducer, acoustic amplitude (pressure) are reached maximum, now known as the "resonance". Between adjacent maxima, the distance variation between the two transducer for $\lambda/2$.

By antinode (or nodes) conditions, the two adjacent antinode (or nodes) the distance between the $\lambda/2$, when the distance L between T1 and T2 equals the integer times of half wave, or when:

$$L = n \cdot \frac{\lambda}{2}(n = 0, 1, 2, 3, \cdots) \qquad (11-6)$$

Receiving transducer T2 to receive sound pressure is great value, on the oscilloscope observation of T2 is converted into electrical signals is maximum.

Because of diffraction and other losses, from left to right the maximum amplitude increases with the distance between T2 and T1 decreases gradually. To measure the wavelength of sound wave, we may change the distance L between T2 to T1 continuously, at this point can be observed in the oscilloscope display signal amplitude by a great change to the minimax ······ such a cyclical change, at the same time, the maximum amplitude decreases. As shown in Fig. 11 – 4, with the signal amplitude of each cyclical change, the distance between T1 and T2 L has been changed $\frac{\lambda}{2}[\Delta L = L_{n+1} - L_n = (n+1)\frac{\lambda}{2} - n\frac{\lambda}{2} = \frac{\lambda}{2}]$, The distance change value can be read by vernier caliper. To calculate the wavelength λ, and then by (11 – 4) and the measured frequency f to calculate f sound v.

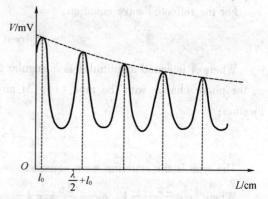

Fig. 11 – 4 Principle of measurement

Due to acoustic amplitude near the maximum along with the change of L is sharp, and near the minimum is flat, so the determination of the location of the maximum more precisely. In addition, because there is a natural frequency of piezoelectric transducer itself, when coupled with the forced vibration frequency is equal to the natural frequency, the piezoelectric transducer will produce resonance. This case vibration amplitude, the largest of the amplitude of sound waves.

So the experiment should be carefully adjust the working frequency of signal generator, the received signal amplitude to the maximum.

(4) Phase comparison method (wave) measure the wavelength of sound waves

Phase comparison method was used to measure sound wavelengths of experiment device as shown in Fig. 11 – 5.

When transmitting transducer and receiving transducer for L, the distance between the transmitting transducer drive sine signal and sine signal receiving transducer receives will be between the phase difference of $\varphi = 2\pi L/\lambda = 2n\pi + \Delta\varphi$.

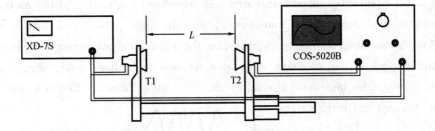

Fig. 11 – 5 Phase comparison method (wave) measuring acoustic wavelength of experimental apparatus

If the transmitting transducer drive sine signal and sine signal receiving transducer receives respectively connected to the X and Y input terminal of the oscillograph, the perpendicular will interfere with the same frequency of sine wave, the synthetic trajectory is called Lissajous figure, as shown in Fig. 11 –6.

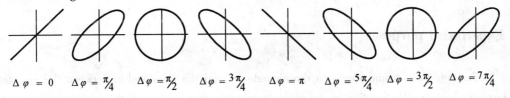

Fig. 11 – 6 Lissajous figure for different phase

Changes when the receiving transducer and the distance between transmitting transducer is equal to a wavelength, the phase difference between transmit and receive signal also just change one period ($\Delta\varphi = 2\pi$), the same graphic will appear. Conversely, when the accurate observation of phase difference change a period when the receiving transducer moving distance, can obtain the corresponding acoustic wavelength lambda, and then based on the frequency of sound waves, the propagation velocity of sound waves.

(5) The time difference method for measuring sound velocity

Time difference method to measure the speed of sound is a commonly used method in engineering application, for example on the ocean ships use sonar to measure the target distance and azimuth, sounder principle of application in water conservancy engineering is the time difference method. Time difference method, the experiment of measuring sound velocity device as shown in Fig. 11 – 3. Sound velocity is measured with method of time difference, need to source can launch a modulated sine pulse signal. Measurement, set the source function selection, with a sinusoidal signal input to the ultrasonic transmitting transducer modulation (fixed on the vernier caliper main ruler of piezoelectric transducer), the ultrasonic pulse, after time t reach the distance L of ultrasonic receiving transducer (on the vernier caliper cursor with piezoelectric ceramic transducer). After receiving transducer receives the pulse signal, energy gradually accumulate, amplitude increase, gradually after the pulse signal. Receiver for damping oscillation, transmit and receive signal as shown in Fig. 11 – 7. Time t can be used the oscilloscope to measure. The distance between the transmitting transducer and receiving transducer by ultrasonic experiment device of vernier caliper to measure, can be calculated by the empirical formula $v = L/t$, the speed of sound.

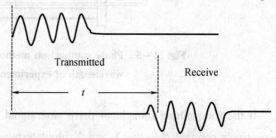

Fig. 11 – 7   Transmit and receive signal

3. Experiment purposes

Understanding of sound waves in the air velocity and the relationship between the gas state parameter; learning the function of the piezoelectric transducer, deepen the understanding of the in sweep vibration synthesis theory; mastering the use of the oscilloscope; and a method of measuring sound velocity in the air.

4. Experiment instruments

ZKY – SSA ultrasonic sound velocity meter, XD – 7 s low frequency signal generator and COS5020B general oscillograph. Now it is introduced as follows:

(1) ZKY – SSA Ultrasonic sound velocity meter

ZKY – SSA Ultrasonic sound velocity meter consists of stents, vernier caliper and two ultrasonic piezoelectric transducers. Two transducer position respectively to the main ruler and vernier relative positioning of the vernier caliper, so relative distance variation between two transducers can be read directly by the vernier caliper. Two transducer has the same structure, which is located in the instrument on the left side of the fixed on the caliper of a transducer, the flat end face to launch sound waves, namely the realization of electroacoustic transformation, the other one is located in the instrument right above the cursor and fixed the transducer, the flat end for receiving and reflected sound waves, realize the acoustoelectric conversion.

ZKY - SSA ultrasonic sound velocity meter on the resonance frequency of the transducer is 30 ~ 40 kHz range; Effective measuring distance is 250 mm; Distance reading accuracy to 0.02 mm.

(2) XD - 7 s low frequency signal generator

XD - 7 s low frequency signal generator panel as shown in Fig. 11 - 8:

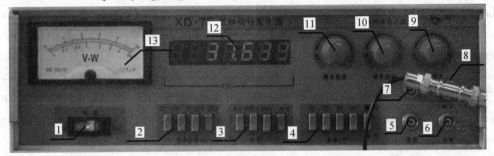

Fig. 11 - 8 XD - 7s low-frequency signal generator panel and the key function

1—the power switch; 2—frequency selection; 3—output power choice; 4—attenuation choice; 5—the signal input terminal; 6—the power output end; 7—TTL level output; 8—voltage output; 9—output amplitude adjustment; 10—frequency fine tuning; 11—frequency rough adjustment; 12—frequency digital display; 13—the output voltage/power meter

(3) General oscilloscope COS5020B type

COS5020B type general oscilloscope panel as shown in Fig. 11 - 9. Where each key (button) function as follows:

Instrument panel number: 1. The calibration signal output terminal; 2. The power indicator light; 3. The power switch; 4. The luminance adjustment, namely brightness adjustment curve; 5. The focus adjustment, or adjust the curve to the clear; 6. Light trace rotation, that is, adjust the horizontal scan lines, parallel to that of the scale line; 7. Brightness scale, namely to adjust the brightness of the scale; 8. CH1 vertical displacement of the channel, namely the adjustment curve of the vertical position; 9. CH1 channel input signal and the vertical amplifier connection mode choice, from top to bottom, DC to AC, respectively; 10. CH1 channel (vertical) signal input terminal; 11. (vertical) CH1 channel signal attenuation, from 5 mV/cm to 5 V/cm is divided into 10 block; 12. (vertical) CH1 channel signal attenuation trimming; 13. The way to choose the CH1 vertical system, and the CH1 channel work alone. Select the "ALT", then the

CH1 and CH2 two channel work alternately, suitable for high speed scan. Select "CHOP", the 250 kHz frequency in turn display CH1 and CH2, scanning is suitable for low speed condition. Choose "ADD", is used to measure the CH1 and CH2 algebra and (CH1 and CH2), if the CH2 fine-tuning knob out, measuring the difference between the two channels; Select CH2, CH2 channel work alone; 14. The oscilloscope earthing terminal; 15. The trigger signal source selection switch, that is, when the trigger signal source switch 25 are effective in the "inside", including "CH1 $(x-y)$" is the CH1 input signal for the trigger signal, in the $x-y$ working status, the signal connection on the X axis. "CH2", CH2 will signal as a trigger signal. "VERT MODE", it is the signal on the screen as a trigger signal; 16. CH2 channel (vertical) signal attenuation, from 5 mV/cm to 5 V/cm is divided into 10 block; 17. CH2 channel (vertical) signal attenuation trimming; 18. CH2 channel (vertical) signal input terminal; 19. The CH2 vertical amplifier channel input signal and connection mode choice, from top to bottom, DC to AC, respectively; 20. The CH2 vertical displacement of the channel; 21. The outer knob to "release time suppression" regulation, the inner regulation knob to "trigger level"; 22. The external trigger signal input terminal; 23. The trigger signal polarity selection; 24. The trigger signal coupling mode selection, "AC" is through the exchange coupling exert trigger signal, the "group" also as the exchange coupling, but there are additional inhibition above 50 kHz signal feature, select "TV" trigger signal synchronous separation circuit should be taken from the TV, the scanning time base selection button 28 should be placed on a TV. "V" or "TV. H" status; "DC" is through the DC coupling exert signal; 25. The trigger signal source selection, "inside" is the internal signal as a trigger signal, trigger signal source within the select key 15 effective at this time. Select the "power" is based on the ac power as a trigger signal. Select "outside" is from the external trigger signal input signals to trigger the 22; 26. The ready light in a single scan; 27. Scan mode selection, of which the "automatic" (AUTO) is a trigger signal to join or trigger signal frequency is lower than 50 Hz while you work, the scanning for the self-excitation method. "Normal" (NORM) is when no trigger signal to join, is in the ready state scan, no line on the screen, below 50 Hz frequency signal is mainly used for observation. "SINGLE" (SINGLE) is used to start, in a SINGLE scan when scanning way has not been three key press, circuit that is in the state, in a SINGLE scan "readiness in a SINGLE scan light is 26" at this time. After press the three scanning way to select one of the key, scanning circuit reset, "readiness in a single scan light is 26" destroyed; 28. (horizontal) Scanning time base selection; 29. (horizontal) Scan fine tune; 30. The curve of horizontal position adjustment.

Measurement of Sound Velocity — Experiment 11

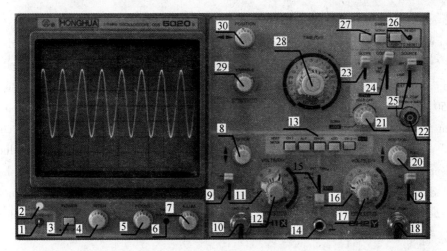

Fig. 11−9　COS5020B general oscilloscope panel and the key function

5. Experiment content and operation key points

Familiar with the instrument

According to the Fig. 11 − 3 connected lines, control instrument read "experimental apparatus" part of the content.

The adjustment of the instrument

(1) The receiving transducer T2 to transmitting transducer and pay attention to leave about 1 cm clearance (to prevent damage to the transducer).

(2) The low frequency signal generator "to select key output power of 3" in the right side "voltage" key press. The output signal cable joint access "8" voltage output interface; The low-frequency signal generator "frequency selective 2" in "20 k to 200 k" gears (fourth from left to right several key), "9" to adjust output amplitude to counterclockwise to output the minimum (to protect the instrument will not be damaged, something is wrong electrical equipment should be such operation, after waiting for instrument response is normal, then according to the need to appropriately increase output). Open the power switch "1", to adjust the frequency coarse adjustment knob "11" to adjust the frequency fine tuning knob "10", the "twelve" frequency digital display shows the frequency of reading about the resonance frequency of the transducer on

159

the ultrasonic sound velocity meter range.

(3) Will receive the output of the transducer T2 signal oscilloscope access to the "CH1 channel (vertical) signal input terminal 10". The oscilloscope vertical system work mode choice "13" in "CH1" position, "CH1 channel (vertical) signal attenuation 11" adjustment to 20 mV/cm or 50 mV/cm position, "(horizontal) scan select 28" adjustment to the position of 10 us/cm, "CH1 channel input signal and the vertical amplifier connection methods to choose 9" to "AC" gear. On of the "power switch 3" oscilloscope, proper adjustment of luminance adjustment 4, "focus on regulating 5", "CH1 channel 8" vertical displacement and horizontal position adjustment "30" curve, and appropriately adjust "9" to adjust output amplitude of the signal generator and power output option 3, choose 4 attenuation, can get the curve of an input signal.

The determination of resonance state

Using oscilloscope to observe received by T2 translates into electrical signals. Adjust the signal generator frequency fine tuning knob "10" (when it is necessary to adjust frequency coarse adjustment knob "11"), which maximizes the oscilloscope display of waveform amplitude. The working frequency of the signal generator is the natural frequency of the transducer.

The determination of T2 maximum starting position

To move slowly T2, can be observed on the oscilloscope waveform amplitude changes. Maximize T2 will be moved to a certain amplitude, fixed T2, record the T2 corresponding gauge readings as the first maximum location data S1, for measuring distance L between T2 and T1 starting position.

Note: in the process of complete operating, shall, from time to time to adjust the oscilloscope "CH1 channel (vertical) signal attenuation 11", it is important to control oscilloscope waveform is displayed on the two peaks are within the scope of the screen, otherwise can't determine whether waveform to maximize.

Using interferometry measuring acoustic wavelength

In order to improve the accuracy of, make full use of data resources, this experiment adopts data processing through "gradual deduction method". To do this, you need two sets of data, make the difference between the two corresponding to each other items in a set of data of 20 half wavelength. Specific means is: slow moving to the right T2, record the first order to appear the oscilloscope waveform of T2 before 10 position reading $S_1, S_2, \cdots, S_{10}$, and then continue to move to the right to T2, and silently count to a maximum 21, began to record the second group, a total of 10 consecutive data $S_{21}, S_{22}, \cdots, S_{30}$. Will the 20 data corresponding to subtract $(S_{21} - S_1)$, $(S_{22} - S_2), \cdots, (S_{30} - S_{10})$, notice that the size 10 difference in theory should be $20 \times \lambda/2 = 10\lambda$, the 10 value addition arithmetic average value, in addition to the 10 to 10λ after it except 10λ. With the frequency f plug type sound velocity v can be obtained though formula (11-4).

Use the phase comparison method (wave) measuring acoustic wavelength

(Use another method to verify measurements, data table from) The source of the output signal with BNC tee joint (T) is divided into two road, all the way used to drive the transducer, the receiving transducer T1, the other all the way to access the oscilloscope "CH2 channel 18" (vertical) signal input, the output of the T2 signal lines still access the oscilloscope "CH1 channel (vertical) signal input terminal 10", the adjustment signal source frequency adjustment knob to transducer resonance state, to adjust the oscilloscope channel CH1 and CH2 channel respectively (vertical) signal attenuation adjustment knob 11 and 16, the two signal similar proportion size (two similar signal waveform amplitude). The above work is completed, press the oscilloscope vertical system work mode selection button in the 13 "CH2 $(x-y)$" button, adjust the time base (horizontal) scanning from 28th to choose knob "$(x-y)$" state, the oscilloscope signal source select key 25 under "INT" $(x-y)$ state, trigger source choice within scope key 15 in "CH1 $(x-y)$", the transmitter drive sine signal with the receiver to receive the sine signal respectively received the X and Y input terminal of the oscillograph, the oscilloscope is in a state of "$(x-y)$", the perpendicular with frequency sine wave interference, the synthesis path is called Lissajous figure, as shown in Fig. 11-6, According to lee, as shown in figure of periodic change measuring wavelength, using data processing through "gradual deduction method". Need to measure the two groups of data, corresponding to each other in the two groups of data between 10 and a half wavelength. Specific means is: slow moving to the right T2, first order to record the oscilloscope in the same state of Mr Lee as shown in figure (preferably a direction of linear state) 10 top location of T2 readings $S_1, S_2, \cdots, S_{10}$, and then continue to move to the right to T2, the second group, a total of 10 continuous recording data $S_{11}, S_{12}, \cdots, S_{20}$. Will the 20 data corresponding to subtract $(S_{11} - S_1), (S_{12} - S_2), \cdots, (S_{20} - S_{10})$, notice the size 10 difference in theory should be 10λ (why?), the same will be after the 10 values in addition to the 10 to 10 lambda arithmetic average value, it in addition to the 10λ. With the frequency f plug type sound velocity v can be obtained though formula (11-4). Their own design appropriate data record form of data processing.

Record the indoor temperature

Record the experiment when indoor temperature t, the expectations theory is used to calculate the velocity of sound, for comparison with the experimental results.

Matters needing attention

(1) The experimental process to keep the work frequency of the signal generator is always working on the natural frequency of the transducer, and maintain a constant output voltage.

(2) Can be adjusted repeatedly the handwheel to determine T2 in precise location near the maximum position, hand bye cheng readings to no effect (why? / why not?).

6. Data recording and processing

Frequency measurement

$f = +/- 0.05$ kHz; $\Delta f = 0.05$ kHz

The wavelength measurement

Room temperature $t = $ ℃; 10 wavelength instrument error Δ instrument $= 0.02$ mm (see Table 11 - 1).

Table 11 - 1　Data record form

	S_1	S_2	S_3	S_4	S_5	S_6	S_7	S_8	S_9	S_{10}
S_i/mm										
	S_{21}	S_{22}	S_{23}	S_{24}	S_{25}	S_{26}	S_{27}	S_{28}	S_{29}	S_{30}
S_j/mm										
S_{j-i}/mm										
$\overline{S_{j-i}}$/mm										
ΔS_{j-i}/mm										
$(\Delta S_{j-i})^2$/mm^2										

Measurement of Sound Velocity — Experiment 11

(1) Class A uncertainty of 10λ: $\Delta A_{10\lambda} = \sqrt{\dfrac{\sum_{i=1}^{n}(x_i-\bar{x})^2}{10-1}} = \sqrt{\dfrac{\sum_{i=1,j=21}^{10,30}(\Delta S_{j-i})^2}{10-1}} = \qquad$ mm

(2) Class B uncertainty of 10λ: $\Delta B_{10\lambda} = \sqrt{\Delta_{\text{仪}}^2 + \Delta_{\text{仪}}^2} = \qquad$ mm

(3) Total uncertainty of 10λ: $\Delta_{10\lambda} = \sqrt{\Delta A_{10\lambda}^2 + \Delta B_{10\lambda}^2} = \qquad$ mm

(4) $\lambda \pm \Delta\lambda = (S_{j-i} \pm \Delta_{10\lambda})/10 = \qquad \pm \qquad$ mm

To calculate the wave velocity value

(1) $v = f \cdot \lambda = \qquad$ m/s

(2) $E_r = \sqrt{\left(\dfrac{\Delta f}{f}\right)^2 + \left(\dfrac{\Delta\lambda}{\lambda}\right)^2} = \qquad$ %

(3) $\Delta v = v \cdot E_r = \qquad$ m/s

(4) $v = \bar{v} \pm \Delta v = \qquad \pm \qquad$ m/s

Compared with the theoretical expectations

(1) $v_{\text{theory}} = v_0\sqrt{1+\dfrac{t}{T_0}} = 331.45\sqrt{1+\dfrac{(16.5)}{T_0}} = \qquad$ m/s

(2) $\Delta'v = |v_{\text{theory}} - v| = \qquad$ m/s

(3) $E'_r = \Delta'v/v_{\text{theory}} \times 100\% = \qquad$ %

7. Analysis and questions

(1) Why did the experiment in transducer resonant condition determine the velocity of sound in air?

(2) Why keep transducer in experimental emission surface parallel to accept below?

(3) Why changes the frequency of the signal source in the experiment?

8. The appendix

The working principle of oscilloscope

The structure of the electron beam oscilloscope tube (CRT) and its effect:

Oscilloscope tube is the core component of oscilloscope, it can make the transition from the electrical signal to light signal, the waveform of the signal under test is displayed by its. Its basic structure as shown in Fig. 11 - 10. It is composed of electron gun, deflection system and screen of three parts. Oscilloscope tube is the function of the speed of the electron beam under the action of deflection system to screen, make the screen to produce bright line. If we can make the measured voltage and scanning electron beam voltage under the joint action of exercise, can make the window on screen form a moving trajectory, in order to display the measured signal waveform. Under the control of the process as the pen in hand draw on paper: electron beam ACTS as the role of the pen, the screen is the role of the paper, the deflection system is the only drawing hand.

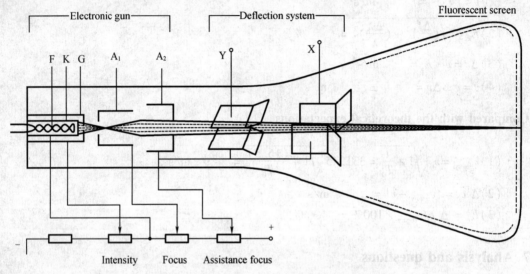

Fig. 11 - 10 Oscilloscope tube structure

F—filament; K—cathode; G—control grid; A1—first anode; A2—the second anode;
Y—vertical deflection plate; X—horizontal deflection plate

(1) Electron gun

Gun grid is controlled by Filament, cathode K, F control grid G, the first speed up the second accelerating anode anode A_1 and A_2 of five parts. Cylindrical cathode filament by using heat. Coated with oxide cathode is a surface of metal cylinder, after being heated surface fire hot electron in large quantities. Control gate is a top have a hole (about 1 mm in diameter) of metal cylinder, it set on the outside of the cathode, its potential than the cathode is slightly lower, hot electron flow in the anode under the accelerated towards the screen. Oscilloscope on the front panel "brightness" adjustment is by adjusting the grid potential in control at the screen of electron density, which changes the screen brightness of light. Anode potential than the cathode potential is much higher, electronic by the electric field between them to accelerate the formation of rays. When the control gate, between the first and second anode potential adjustment when appropriate, electron gun of the electric field of electron beam focusing effect, so the first anode is also called the focus anode. The second anode potential higher, also known as accelerating anode. Panel "focus" on the adjustment, is to adjust the anode potential made the light spot on the screen become bright, clear dots. Some oscilloscope and "auxiliary focus", is actually the second anode potential adjustment.

(2) Electric deflection system

It is composed of two mutually perpendicular deflection plate, a pair of vertical deflection plate $Y - Y$, a pair of horizontal deflection plate $X - X$. Plus dozens of voltage between the deflection plate, electron beam by, its movement direction to deflect, thus able to control the electron beam bombardment point on the screen (spot).

(3) Screen

The screen coated with phosphor, electronic beat up it's light, form a flare. Different materials and different proportion of phosphor luminescence color is different, the continuation of light-emitting time (generally called afterglow time) is also different. For DC or low frequency signal display appropriate uses long afterglow screen. There is a piece of transparent, in front of the screen with the coordinates of the board, the position of light spot. In good performance oscilloscope, the scale directly on screen in the glass surface, with the fluorescent powder is clingy together to eliminate the parallax, light spot position can be measured more accurate.

X axis and Y axis voltage amplification (and attenuation) device

Oscilloscope at least X and Y the two input channels, each has its own, independent voltage amplifier. Because of the small signal voltage under test, usually using the Y axis voltage amplifier amplification, magnification ratio by the gain control knob to adjust (adjustment can be stepping on fine-tuning forms, can also be a continuous form). The above is through "V/div" deflection sensitivity selector switch. The large signal voltage eventually add corresponding deflection in the oscilloscope tube plate, used to control of the fluorescent light blasted off hot electron in the corresponding direction of displacement. Displacement law completely the same as the voltage variation law of the input signal, the light will be along the left and right along with the change of signal voltage (or up and down) for cycloid movement, form a horizontal or vertical line. If the signal voltage is too large, generally at the input to the input voltage signal through precision attenuator attenuation to the appropriate size, then send voltage amplifier amplification.

As shown in Fig. 11 - 11 signal to Y input, screen without scanning signal line images as shown in Fig. 11 - 12.

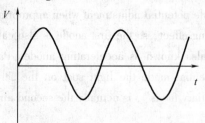

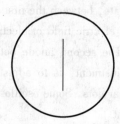

Fig. 11 - 11   Signal under test

Fig. 11 - 12   Without scanning signal oscilloscope display waveform

Sweep signal generator

To see on the screen for the waveform of voltage signal, this signal must be a voltage on the Y deflection plate, at the same time in the X deflection plate and a periodic voltage linear change over time, the voltage linear with time to rise to the maximum to rapidly after starting value (zero), after repeated, as shown in Fig. 11 - 13. Because of the voltage and time curve such

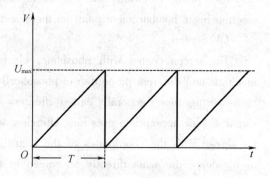

Fig. 11 - 13   Repeat signal

as blade, so called the sawtooth wave. Generated sawtooth voltage of the device is called "sawtooth wave generator". Through the oscilloscope control panel on the knob, however, can only add the sawtooth wave signal to the horizontal deflection plate, at this time if no signal on the Y axis, electron beam on screen along the horizontal direction only a limited range of linear motion, a level of light will flash on the screen. If the sawtooth wave frequency is low, you can see a bright spot again, moving from left to right, this is scanning the origin of the word.

Waveform display

Plus a Y deflection plate at any time, intercropping sine voltage if X deflection plate voltage is zero, the screen will have a vertical bright line; If in the X deflection plate at the same time and add a sawtooth voltage is proportional to the time, and the sawtooth voltage cycle and Y deflection plate signal voltage of the same cycle, the screen shows as shown in Fig. 11 - 14 full waveform of a cycle.

Voltage U_x and U_y are zero in start point a, screen window at A, time from a to b, the only function voltage U_y, highlight the displacement along the vertical direction for bB_y, screen window in B_y place, and at the same time after joining the U_x, electron beam subjected to U_y upward deflection, both at the same time subjected to the U_x deflection to the right (window horizontal displacement bB_x), thus highlights not B_y place, and at B. Over time, and so on, can show the sinusoidal waveform. So, see on the screen of sine curve is actually two perpendicular motion synthesis.

Due to the sawtooth voltage is a cyclical change (back and forth along the horizontal direction of the electron beam scanning), makes on Y deflection signal waveform is repeated. And since the cycle of two signals is equal ($T_x = T_y$), each scanning waveform coincide, and screen display of the waveform is stable. Cycle of change sawtooth wave cycle, if it is Y deflection plate signal voltage cycle 2 or 3 times, the screen display 2 or 3 cycles Y deflection plate signal waveform. When Y signal voltage cycle T_y and X axis sawtooth voltage cycle T_x meet relationship ($T_x = nT_y$), screen display waveform on the number of cycles of n stable waveform.

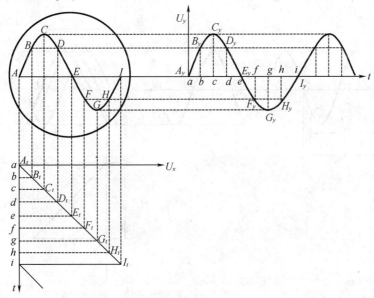

Fig. 11 – 14 Oscilloscope display waveform principle

T_y is actually determined by the voltage being measured, T_x is determined by oscilloscope in the sawtooth wave generator, the two unrelated to each other. Although adjustable sawtooth "scan" and scan "fine-tuning" the $T_x = nT_y$, but as a result of the T_x and T_y come from two different systems, inevitably, each change in the experimental process, resulting in waveform is not stable, when the T_x slightly less than nT_y, waveform will move to the right side of the screen, move speed depends on the size of the difference between T_x and nT_y. Similar, when T_x slightly greater than nT_y, waveform will move to the left side of the screen, for example, the sawtooth voltage cycle T_x than the cycle of the sine wave voltage being measured T_y smaller, $T_x : nT_y = 7 : 8$. As shown in Fig. 11 – 15, in the first scan cycle, screen displayed on the sine signal curve segment between 0 to 4 points, the starting point at the O; In the second period, showed that the curve between four to eight points, starting at 4; The third period showed that the curve between 8 and 12 points, starting at 8. This waveform is displayed on the screen every time don't overlap, caused the phenomenon of waveform to the right. Likewise, if the T_x slightly bigger than a nT_y, would cause the waveform to the phenomenon. Described above phenomenon in use process will often appear in the oscilloscope. Investigate its reason is scanning the cycle of the control voltage and the measured signal cycle no integer multiple relationship, so that each scan at the beginning of the starting point of the waveform curve are not in the same place.

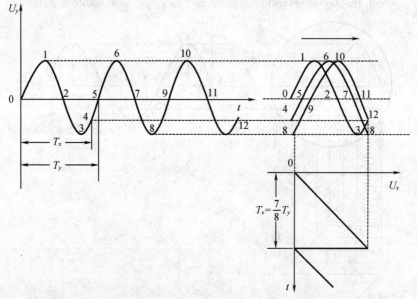

Fig. 11 – 15 The T_x and T_y no integer times when the movement of the waveform

In order to observe the stability of the signals to be measured waveform, can take a whole step method, namely gave sawtooth wave generator under test signal sampling, forcing the sawtooth wave frequency following changes T_y of signal under test cycle, which can guarantee the

$T_x = nT_y$ set up all the time, screen can always get a stable waveform.

Discussion by above knowable, oscilloscope display stable waveform is: the condition of signal under test has large enough strength, at the same time sawtooth scanning signal cycle should be kept for the integer times of signal under test cycle.

Measurement of Metal Electronic Work Function

1. Background and application

The phenomenon of electron emitting from hot-metal is called thermo-electron emission. One of the purposes of researching thermo-electron emission is to select appropriate cathode substance. Experiment and theory prove that the main parameter affecting filament emission current density is filament temperature and work function of filament substance. The higher the temperature of filament is, the higher the emission current density is; therefore, ideal thermo-electron emission body of simple metal must have less work function and have higher fusion point to make working temperature increase to obtain higher emission current. Since thermo-electron emission depends on the work function and temperature of the material, the material with high fusion and low work function should be selected as the cathode. Nowadays the most widely used simple metal is tungsten, and the experiment uses Richardson's straight line method to measure the work function of tungsten and then the basic law of thermo-electron emission is deeply understood.

Owen Willianms Richardson (see Fig. 12 – 1) is the founder of "Richardson law". During World War II, he devoted himself to studying radar, sonar, electronics experimental instrument, magnetron and klystron, etc.

Fig. 12 – 1 Richardson
(1879—1959)

Measurement of Metal Electronic Work Function — Experiment 12

On November 25, 1901, Richardson read his thesis in Cambridge Philosophy Institute, saying that if heat-radiation was produced by the particles emitted by the metal, saturation current should obey the following law:

$$I = AT^{\frac{1}{2}} \exp\left(-\frac{b}{T}\right)$$

The law has been completely proved by the experiments, at that time Richardson, who was only 22, amazingly laid the base to win Nobel Physics Prize 27 years later. In 1911, Richardson put forward thermo-electron emission formula (Richardson Law) that has experienced the trial of quantum mechanics of 1920s:

$$I = AST^2 \exp\left(-\frac{e\varphi}{kT}\right)$$

Richards won the Nobel Physics Prize of 1928 because he achieved great success in researching thermion phenomenon, especially finding out Richardson Law. Richards's thermo-electron emission theory established a solid foundation for radio development (see Fig. 12 - 2), and now a lot of cathodes of electric vacuum devices work in terms of thermo-electron emission (see Fig. 12 - 3).

Fig. 12 - 2 Wireless antenna equipment   Fig. 12 - 3 Vacuum valve

2. Experiment principles

In hard valves, a cathode K made by metallic wire to be tested is heated by current I_f. When a positive voltage relative to cathode is added to another anode, a current passes through the outer circuit connecting the two electric poles as shown in Fig. 12 - 4, and the phenomenon is called thermo-electron emission. The work function of cathode material can be measured by studying thermo-electron emission law to select appropriate cathode materials.

(1) The work function of electrons

In accordance with metallic electron theory of solid physics, the conduction electrons of metal have certain energy, but are in degenerate state: each level of single atom splits apart into many very adjacent energy levers, just like a continuous band, called "energy band". Modern electronic theory thinks electronic energy in metal does not follow Maxwell distribution, but Fermi-Dirac statistic formula, i. e.

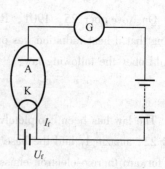

Fig. 12 – 4 Schematic diagram

$$f(W) = \frac{dN}{dW} = \frac{4\pi}{h^3}(2m)^{3/2}W^{1/2}\frac{1}{\exp(\frac{W-W_i}{kT})+1} \qquad (12-1)$$

At the temperature of absolute zero, its energy distribution curve is shown by Curve 1 in Fig. 12 – 5; at the moment, the maximum energy the electron has is Fermi level W_i. When the temperature rises, energy distribution curve of electron is shown by Curve 2 in Fig. 12 – 5, in which a few electrons with higher energy have energy higher than W_i and the number of such electron decreases with the increase of energy progressively to index law.

At general temperature, why is there almost no electron struggling to come out from its surface in metal? The reason is that on the metal surface there is an even charge layer of positive charge with the thickness about 10^{-10} m to prevent electrons escaping from metal surface. That is to say there is barrier potential of position energy W_a between metal surface and outside. In order to escape from metal, at least electrons have to the energy of W_a, i. e. the resistance of even charge must be overcome (see Fig. 12 – 5). At the temperature of absolute zero, the minimum energy required for electron escaping from metal is called electron work function, which is expressed with W_0, and obviously there is

$$W_0 = W_a - W_i = e\varphi \qquad (12-2)$$

Common unit of W_0 is electron-volt (eV), which shows that the metal at the temperature of absolute zero is made to have the energy required to be given for the electron with maximum energy W_i escaping from metal surface; φ is called escape unit, whose value equals to electron work function expressed with electron-volt, and the unit is volt (V).

(2) Thermo-electron emission formula

In accordance with energy distribution formula (12 – 1) of Fermi-Dirac, Richardson-Dushman formula of thermo-electron emission can be derived, i. e.

$$I = AST^2 e^{-\frac{e\varphi}{kT}} \qquad (12-3)$$

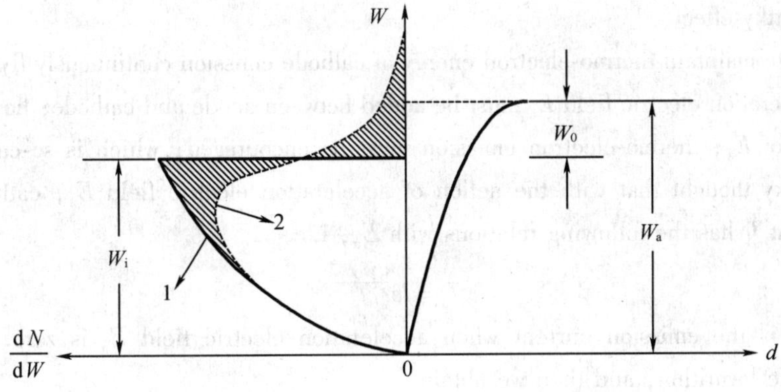

Fig. 12-5 Electron energy distribution curve

Where I is current intensity of thermo-electron emission, and the unit is (A); S is effective emission area of cathode metal, and the unit is (cm^2); T is the absolute temperature of hot cathode, and the unit is (K); $e\varphi$ is the electron work function of cathode metal, and the unit is (eV); K is Boltzmann's constant, and $k = 1.38 \times 10^{-23}$ J/K; A is coefficient related to cathode chemical purity.

In principle, we can calculate cathode work function $e\varphi$ according to formula (12-3) as long as I, A, S and T are measured, but the difficulty is to measure A and S; therefore, in actual measurement, the so-called Richardson's straight line method is usually used.

(3) Richardson's straight line method

Both sides of formula (12-3) are divided by T^2, then take logarithm, and we obtain

$$\lg \frac{I}{T^2} = \lg(AS) - \frac{e\varphi}{2.303kT} = \lg(AS) - 5040\varphi \frac{1}{T} \qquad (12-4)$$

From formula (12-4), $\lg \frac{I}{T^2}$ and $\frac{1}{T}$ have a linear relation. If we take $\lg \frac{I}{T^2}$ as the longitudinal coordinate, and $\frac{1}{T}$ as the horizontal coordinate to graph, then electron escape unit φ can be found out from the obtained slope of straight line, which is called Richardson's straight line method. Its merit is that the specific values of A and S can not be necessarily found out, φ value can be obtained directly from T and I, and the effect of A and S is only to make $\lg \frac{I}{T^2} - \frac{1}{T}$ move parallel in straight line. The method is widely used in experiments, scientific researches and production.

(4) Schottky effect

In order to maintain thermo-electron energy of cathode emission continuously flying to anode, an extra acceleration electric field E_a must be added between anode and cathode; however, due to the existence of E_a, thermo-electron emission must be encouraged, which is so-called Schottky effect. Schottky thought that with the action of acceleration electric field E_a, cathode emission electric current I_a has the following relations with E_a, i.e.

$$I_a = Ie^{\frac{4.39\sqrt{E_a}}{T}} \qquad (12-5)$$

Where I is the emission current when acceleration electric field E_a is zero. For formula (12-5), take logarithm, and then we obtain

$$\lg I_a = \lg I + \frac{4.39}{2.303T}\sqrt{E_a} \qquad (12-6)$$

If cathode and anode are made into axial cylinder, and ignore the touch potential difference and other influences, the acceleration electric field can be expressed as

$$E_a = \frac{U_a}{r_1 \ln \frac{r_2}{r_1}} \qquad (12-7)$$

Where r_1 and r_2 are the radius of cathode and anode, respectively; U_a is acceleration electric voltage. Substitute formula (12-7) into formula (12-6), and then we have

$$\lg I_a = \lg I + \frac{4.39}{2.303T} \frac{1}{\sqrt{r_1 \ln \frac{r_2}{r_1}}} \sqrt{U_a} \qquad (12-8)$$

Known from formula (12-8), at certain temperature and tube structure, $\lg I_a$ and $\sqrt{U_a}$ have a linear relation, so take $\sqrt{U_a}$ as the horizontal coordinate, and $\lg I_a$ as the longitudinal coordinate to make straight lines; the elongation line of straight line is crossed to axis at $\sqrt{U_a} = 0$, and the cross point is $\lg I$. Emission current can be determined at certain temperature when acceleration field is zero (shown in Fig. 12-6).

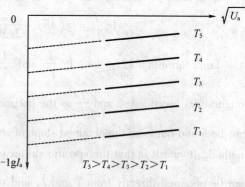

Fig. 12-6 $\sqrt{U_a} - \lg I_a$ curve

(5) The measurement of temperature

It can be seen from thermo-electron emission formula that filament temperature T has a great influence on emission current, so the accurate measurement of temperature is a very important problem. In the experiment, radiated light pyrometer is used to measure filament temperature. In laboratory, the relations between filament heating current and filament temperature are listed in Table 12-1, the corresponding filament temperature to the readings of filament current can be directly checked out, and it is unnecessary to carry out complicated calculation and measure.

Table 12-1   The relations of ideal diode filament current and temperature

Filament heating current I_f/A	0.500	0.550	0.600	0.650	0.700	0.750	0.800
Filament temperature $T/10^3$ K	1.72	1.80	1.88	1.96	2.04	2.12	2.20

3. Experiment purposes

As many cathodes of electric vacuum devices work through thermo-electron emission nowadays, know the basic law of thermo-electron emission through the experiment; learn to use Richardson's straight line method to measure the electron work function of tungsten spiral; verify Schottky effect.

4. Experiment instruments

The instruments used for the whole set of experiment system includes work function determinator, which is composed of tube diode, current and voltage regulator, and ampere meter and avometer, as shown in Fig. 12-7.

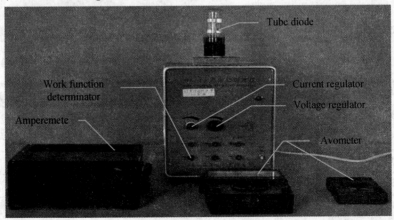

Fig. 12-7   Experiment instruments

5. Experiment content and operation key points

(1) Know the instruments, turn on the electric power and preheat for 10 minutes.

(2) Take filament current from 0.600 ~ 0.750 A, and measure once every 0.050 A. Corresponding to every filament current, add 25 V, 36 V, 49 V, 64 V, 81 V, 100 V, 121 V and 144 V to the anode, measure a group of currents I_a for each voltage, and record the experiment data in Table 12-2.

(3) After converting every quantity in Table 12-2, fill in Table 12-3, and make curves of $\sqrt{U_a} - \lg I_a$, and find out emission current logarithm $\lg I$ (intercept) when acceleration electric field is zero.

(4) Fill the data in Table 12-4, make straight line of $\lg \dfrac{I}{T^2} - \dfrac{1}{T}$, calculate escaping potential and work function of tungsten in accordance with the slope of the straight line, compare it with standard values (4.5 eV) and obtain relative error.

Note:

①Filament is a little fragile, so handle it gently when using, slow down when heating and cooling down, and avoid shaking the filament strongly after filament is red-hot.

②Due to the lag of filament heat balance, preheat for a few minutes; every time the filament current is adjusted, get a group of anode current, and wait a moment for stabilization.

③In order to protect ideal diode, adjust filament heating current lower than 0.6 A before turning off the equipment, and then turn off the electric power.

6. Data recording and processing

Table 12-2 Data recording table

U_a/V		25	36	49	64	81	100	121	144
I_f = 0.600 A	I_a /μA								
I_f = 0.650 A									
I_f = 0.700 A									
I_f = 0.750 A									

Measurement of Metal Electronic Work Function Experiment 12

Table 12-3 Data recording table

T \ lgI_a \ $\sqrt{U_a}$	5.0 V	6.0 V	7.0 V	8.0 V	9.0 V	10.0 V	11.0 V	12.0 V
T = 1 880 K								
T = 1 960 K								
T = 2 040 K								
T = 2 120 K								

Table 12-4 Data recording table

$T/10^3$ K	1.88	1.96	2.04	2.12
lg$I$				
lg$\frac{I}{T^2}$				
$\frac{1}{T}$				

Slope of straight line k = _____,   Escape unit φ = _____(V),
Work function $e\varphi$ = _____(eV),   Relative error E_r = _____%.

7. Analysis and questions

(1) Why can electrons not escape in general conditions, what method is used for electron escape? Explain electron escape principle.

(2) If the filament is not preheated, what influence can be produced on the experiment?

(3) What is Richardson's straight line method, what merit does the method have?

Experiment 13

Michelson Interferometer

1. Background and application

The interference phenomenon of light is a manifestation of the wave property of light. When a light beam is divided into two beams and meet again by passing different paths and the light path difference is less than the coherence length of this beam, the interference phenomenon will occur. Michelson interferometer is a precise optical instrument designed for demonstrating the effect of the hypothetical "aether wind" on the speed of light by the German-born American physicist Michelson in 1881, which, according to amplitude distribution, produces two beams of light to realize the interference phenomenon. Michelson interferometer has the characteristics that the four components, i. e. the light source, two reflectors, and the receiver (the observer), are completely separated in the space, which makes it easy to install other components in the optical path. With its delicate design, easy optical path and high precision, Michelson interferometer is widely used in many fields. Its measurement accuracy was up to 1/400,000,000 in 1887. Since its birth in 1881, Michelson used this device to implement three famous experiments: the Michelson-Morley experiment, the experiment of measuring the fine structure of spectrum and the experiment of calibrating the length unit by utilizing the light wavelength, among which the Michelson-Morley experiment denied the existence of aether and the negative conclusion of this experiment made Einstein to put forward the special theory of relativity in 1905.

Designed for studying the aether and developing for the need of high-tech measuring tool, the principle of Michelson interferometer and its application have become the core of interferometry detection technique, and gradually penetrate into various branches of science and engineering.

Michelson Interferometer   Experiment 13

With this device, people can easily observe various interference phenomena, e. g. equal thickness interference, equal inclination interference, the various movements of fringes, etc.; and do various precise detections, e. g. measuring the wavelength of monochromatic light, the refractive index of transparent bodies (solid, liquid, gas), the coherence length of the light source, etc. Developed from the basic principle of Michelson interferometer, various interferometers, such as the contact interferometer, laser length measuring instrument, Fourier light slitting interferometer (see Fig. 13 – 1) and Twyman interferometer (see Fig. 13 – 2), etc. have been widely used in production and scientific research fields. In a word, Michelson interferometer is the prototype of many modern interferometers.

Fig. 13 – 1 Fourier light splitting interferometer   **Fig. 13 – 2 Twyman-Green interferometer**

Nonlinear Michelson Interferometer

In the so-called nonlinear Michelson interferometer, the plane mirror of one interferometric arm of the standard Michelson interferometer is exchanged for a Gires-Tournois interferometer or Gires-Tournois etalon, and the light field sent from the Gires-Tournois etalon interferes with the reflection light field of the other interferometric arm. Since the phase variation resulted from the Gires-Tournois etalon is related to the wavelength of the light and has a step response, Nonlinear Michelson Interferometer has some special applications, such as interleaver in optical fiber communication, etc. Furthermore, the plane mirrors of the two interferometric arms of the standard Michelson interferometer can both be exchanged for Gires-Tournois etalon; at this time, Nonlinear Michelson interferometer can produce stronger nonlinear effect and can be used to make antisymmetric interleaver.

Optical Fiber Michelson Interferometer

Optical fiber sensor has the characteristics of high sensitivity, corrosion resistance, flame and explosion proof, anti-electromagnetic interference, small size, lightweight and no need of power supply; therefore, it can be used to make optical fiber sensor with diverse adaptabilities. Optical fiber sensors based on the principle of Michelson interferometer have been widely used in the study of pressure, temperature, strain, and refractive index measurement. As for structures and buildings demanding safety and reliability, such as high-rise buildings, bridges, nuclear waste repositories, dams, oil depots, etc., if a Optical fiber sensor of Michelson interferometer can be embedded into the structure to make a smart structure, people can detect the real-time safety situation of the structure, maintain it in time, and greatly reduce the structural safety accidents (see Fig. 13-3).

(a)   (b)   (c)   (d)

Fig. 13-3   Applications of optical fiber Michelson interferometer
(a) Bridge health monitoring; (b) Historic building conservation;
(c) Dam safety monitoring; (d) Offshore exploration activity monitoring

Albert Abraham Michelson (December 19, 1852—May 9, 1931) (see Fig. 13-4), an American physicist, was born in Strzelno, Provinz Posen in the Kingdom of Prussia (now Poland). He moved to the U.S. with his parents in 1855. He graduated from the U.S. Naval Academy in 1873, and was appointed professor and the first head of the department of physics at University of Chicago in 1892. Michelson was a member of the Royal Society, the National Academy of Sciences, the American Physical Society and the American Association for the Advancement of Science. In 1907, Michelson had the honor of being the first American to receive a Nobel Prize in Physics "for his optical precision instruments and the spectroscopic and metrological investigations carried out with their aid".

Fig. 13-4   A. A. Michelson

His first important contribution is the invention of Michelson interferometer and accomplished the Michelson-Morley experiment with it. According to classic physics theory, light and even all electromagnetic waves must be spread via static aether. The earth revolution produces relative motion of aether; therefore, in the two vertical directions of the earth, the time the light passes through the same distance ought to be different, and this difference ought to produce a 0.04 interference fringe in Michelson interferometer. In 1881, Michelson did not observe the interference fringe in the experiment. In 1887, cooperated with the famous chemist Morley, he improved the device, but still did not find it out. The result of this experiment uncovered the defects of aether theory, shook the basis of classic physics, and paved road for the special theory of relativity.

Michelson is the first scientist who advocated using the wavelength as the length standard. In 1892, he used a special interferometer to determine that the wavelength of cadmium red line is 64,384.696 nm, taking the French prototype meter as the standard, on condition that the temperature is 15 ℃ and air pressure is 760 mmHg; and thus one meter is 1,553,164 times of the wavelength of cadmium red line. This was the first time for human beings to obtain one length standard that is permanent and cannot be destroyed.

In terms of spectroscopy, Michelson found out the fine structure of Hydrogen spectrum and the hyperfine structure of mercury and thallium spectrum, which played an important role in modern atom theory. He also utilized the visibility curve method invented by himself to study the relationship between line shape and pressure and the relationship between line broadening and molecular motion, the results of which had great impact on modern molecular physics, atomic spectrum and laser spectroscopy. In 1898, he invented an echelon grating to study Zeeman effect, the resolution power of which greatly outclassed ordinary grating diffraction.

Michelson is an outstanding experiment physicist, the experiments of whom were famous for precise designs and high accuracy. Einstein once praised him as "an artist in the science".

2. Experiment principles

(1) Optical path

Michelson interferometer is a division-of-amplitude dual beam interferometer, and the principle of its optical path is shown in Fig. 13-5. The light beam sent from monochromatic light source S is first divided into two beams, ① and ②, by the half-reflection half-transmission film on the back surface A of the beam splitter G_1, and the light intensity of these two beams are nearly the same and perpendicular. Then beam ① and beam ② are reflected by plane mirror M_1

and M_2, respectively; then are transmitted and reflected by A to become two parallel beams; and finally pass the lens and overlay on the focal plane. One point should be paid attention to, i.e. beam ① reflected by M_1 passes G_1 three times, while beam ② reflected by M_2 only passes G_1 once. In order to compensate the optical path difference, a compensator G_2 is laid on the optical path, the material and thickness of which are the same as G_1, and strictly paralleled to G_1. In this way, when M_1 and M_2 which are strictly symmetrical by the half-reflecting plane A, the optical paths of beam ① and ② are the same regardless of the wavelength of the light. Therefore, it's possible to utilize it to observe white light interference. When calculating the optical path difference of beam ① and ②, the adding of G_2 makes us only consider their geometric distance difference in the air instead of calculating their optical path in the beam splitter.

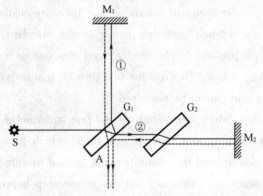

Fig. 13-5 Optical path of Michelson Interferometer

(2) Non-localized interference produced by point light source

Equivalent optical path of non-localized interference is shown in Fig. 13-6, where M_2' is the virtual image reflected by M_2 through plane A. Adjust Michelson interferometer to its standard state to make M_2 and M_2' parallel to each other and make the distance between M_2 and M_2' as d. The point light source S which is focused by convex lens is a strong point light source, whose light reflected by M_1 and M_2 is equal to the coherent light beam emitted by the virtual light source S_1 and S_2', and the distance between S_1 and S_2' is as twice as the distance between M_1 and M_2', namely, $2d$. The spherical waves emitted by virtual light source S_1 and S_2' are coherent everywhere in their space, showing the phenomenon of non-localized interference, and its interference fringe may be the circular ring, the oval ring or the arc interference fringe according to different positions in the space. Usually the viewing-screen F is placed in the position vertical to the line from S_1 to S_2', the distance from the screen to S_2' is R, the interference fringe in the screen is a group of concentric circles whose centre is O.

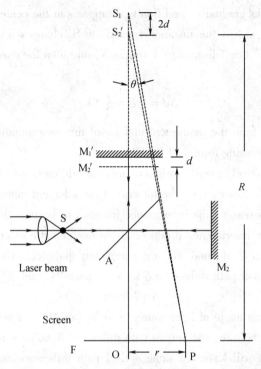

Fig. 13 – 6 The optical path of non-localized interference

Set the distance from S_1 and S_2' to the point P in the viewing screen's optical path difference as δ, and then we have:

$$\delta = \sqrt{(R+2d)^2 + r^2} - \sqrt{R^2 + r^2} = \sqrt{R^2 + r^2}\left(\sqrt{1 + 4(Rd+d^2)/(R^2+r^2)} - 1\right)$$

Under normal circumstances, $R \gg d$, so utilize binomial theorem and ignore high order term of d, and then we have:

$$\delta \approx \sqrt{R^2+r^2} \times \left[\frac{4(Rd+d^2)}{2(R^2+r^2)} - \frac{16R^2d^2}{8(R^2+r^2)^2}\right] = \frac{2dR}{\sqrt{R^2+r^2}}\left[1 + \frac{dr^2}{R(R^2+r^2)}\right]$$

So

$$\delta = 2d\cos\theta\left(1 + \frac{d}{R}\sin^2\theta\right) \qquad (13-1)$$

From formula (13 – 1), we can know that:

①When $\theta = 0$, the optical path difference is maximum, i.e. $\delta = 2d$, which means the interference order corresponding to the centre of the circle is the highest. Move M_1, and if d

increases, we can see the rings burst out from the center one by one and then extend outwards; if d reduces, the ring shrinks gradually, and finally disappears at the center. Once "burst out" (or "disappear") a ring, it means the distance from S_1 to S_2' changes a wavelength (λ). If the number of the "burst out" (or "disappear") ring is N, and then the corresponding mirror M_1 will move Δd:

$$\Delta d = \frac{1}{2}\delta = \frac{1}{2}N\lambda \qquad (13-2)$$

Obviously, read Δd from the instrument and count the corresponding N, the wavelength of light can be calculated by using formula (13 – 2).

②As for the comparatively great d, when optical path difference δ changes a wavelength, the θ's variable quantity will reduce, i.e. the interval of two adjacent round interference fringes will be smaller, so when d increases, the interference fringes will become thinner and more intense.

(3) Equal inclination interference produced by monochromatic extended source

In the Fig. 13 – 6, if M_1 and M_2' are completely parallel, when using monochromatic extended light S, the optical path difference δ between beam ① and ② at point P is:

$$\delta = 2d\cos i \qquad (13-3)$$

where i is the incident angle of light source S to M_1 (or M_2'). From formula (13 – 3), we can know that when d is definite, the optical path difference δ varies with i, the light which have the same incident angle i will have the same optical path difference and overlap at the infinity (The interference will be located at the infinity). If inserting a convex lens (or using the eyes to observe it), we can see the superimposed interference pattern at the focal plane, and this interference is called equal inclination interference and the interference fringe will be a group of concentric circles. If moving back and forth the mirror M_1, we can also see the interference fringes "burst out" (or "disappear") at the centre of the circle. Interference pattern of equal inclination in the different optical path difference is shown in Fig. 13 – 7.

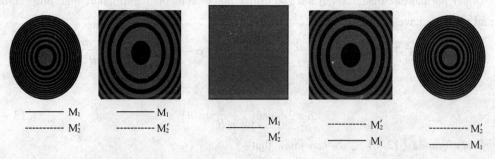

Fig. 13 –7 Equal inclination interference fringe with different optical path differences

(4) Equal thickness interference produced by monochromatic extended light

When the distance between M_1 and M_2' is very close and there is a very small angle, a wedge thin air layer is produced between M_1 and M_2'. At this time, if using monochromatic extended light to irradiate it, we can see the equal thickness interference fringe at the surface of thin air layer, as is shown in Fig. 13 – 8. After light ① and ② emitted by the light source S are reflected by M_2' and M_1, they will overlap at the surface of the thin layer (as M_2' in the Fig. 13 – 9) and interfere with each other. When the angle φ is small, the optical

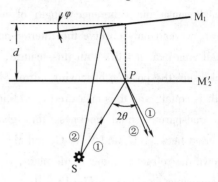

Fig. 13 – 8   The light path of equal thickness interference

path difference between light ① and ② is still approximately equal to $\delta = 2d\cos\theta$, where d is the thickness of the air layer at the point P and φ is the incident angle. At the intersection of M_1 and M_2', $d = 0$; if not considering the additional phase difference produced by mirror reflection, straight bright fringe will appear when $\delta = 0$, which is called central bright fringe.

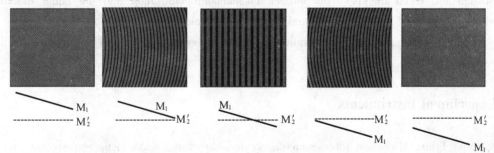

Fig. 13 – 9   Different equal thickness interference fringes with different optical path differences

If the incident angle is not large, $\cos\theta$ can be taken as the approximate value $1 - \frac{1}{2}\theta^2$, so

$$\delta = 2d - d\theta^2 \qquad (13-4)$$

Near the central bright fringe, the interference fringe is the straight fringe which is substantially parallel to the central bright fringe. With the increase of θ, the patterns will become tortuous. The Fig. 13 – 9 is the equal thickness interference image with different optical path differences. From formula (13 – 4), we can know that d must be increased if wanting to maintain the same optical path difference δ, i.e. the bending direction of the interference fringe will be convex to the central bright fringe.

(5) Equal thickness interference produced by white light

Because the coherence length of white light is very short, we can only observe the interference fringe only in a small area near $d = 0$. With this feature, we can accurately make sure the position of moving mirror M_1 when the optical path is equal, and thus measure the additional optical path of the transparent equal thickness thin glass plate which is inserted between G_1 and M_1 or G_2 and M_2. If we have already known the refractive index of the plate, we can calculate the thickness of it, and the experimental phenomenon is shown in Fig. 13 – 10.

Fig. 13 – 10   White light interference

3. Experiment purposes

(1) Understand the working principle of Michelson Interferometer.

(2) Based on mastering how to use and adjust Michelson Interferometer, observe the equal inclination and equal thickness interference of monochromatic light and the equal thickness interference produced by white light, and deeply understand the theory of optical interference.

(3) Learn the principle of determining laser wavelength He-Ne and the method of data processing.

4. Experiment instruments

He-Ne laser, Michelson interferometer, keyhole aperture, short focus convex lens and an incandescent lamp.

(1) He-Ne laser

He-Ne laser is a monochromatic light source with a relatively long coherence length, the wavelength value of which is $\lambda = 632.8$ nm. It composes of two parts: high voltage DC power supply and a laser tube. Usually we use the mini size laser tube whose intra-cavity tube length is 25 cm; mode of output light is single transverse mode; within the coherent length, each spot in cross section of the light is completely coherent, and the output power is 0.5 ~ 3.0 mW. The laser tube whose length is more than 50 cm is the external cavity laser tube with Brewster window, whose output is complete linearly polarized light of single transverse mode and the output optical power is larger than 50 mW.

When no load, the high voltage DC power supply of He-Ne laser outputs voltage up to

10,000 volts, and thus the positive and negative electrode must be connected to the electrode of He-Ne laser tube correctly; thereinto, the negative electrode of the power supply should be connected to the cylindrical Aluminum negative electrode of the laser tube, while the positive electrode should be connected to the positive electrode of the laser tube. If not, the service life of the tube will be much shorter. The electrode with electricity mustn't be touched and short-circuited. Most of the output beam power of He-Ne laser is above 1mW, and thus human eyes mustn't see it directly without the beam expander, for the safety standard of human eyes it must be less than 0.1 mW.

(2) Michelson interferometer

The structure of Michelson interferometer is shown in Fig. 13 – 11.

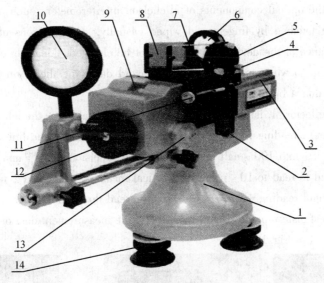

Fig. 13 – 11 The structure of Michelson interferometer

1—base; 2—mutually perpendicular tension spring screw; 3—guide rail; 4—M_2 adjusting screw; 5—stationary reflecting mirror M_2; 6—movable reflecting mirror M_1; 7—compensator G_2; 8—beam splitter G_1; 9—reading window; 10—frosted screen; 11—mutually perpendicular tension spring screw; 12—coarse adjust hand wheel; 13—fine adjustment drum; 14—horizontal adjustment screw

Reflector M_2 is fixed; reflector M_1 is installed on the precise guide rail; with the help of worm and worm gear system, adjust coarse adjustment handwheel 12 and fine adjustment drum 13, and the space position of reflector M_1 can be changed. At the back of reflector M_1 and M_2, there are three adjusting screws (do not over-tighten the screws in the experiment) to adjust the inclination of reflector M_1 and M_2, respectively. Below reflector M_2, there are two mutually perpendicular tension spring screws 2 and 11 attached to reflector M_2 to adjust its inclination much more precisely. Michelson interferometer still has some accessories, such as frosted screen 10, a telescope, etc. which can be easily installed to the device to observe non-localized interference (or equal inclination interference).

Reflector M_1 is driven by worm and worm gear system; therefore, avoiding its idle motion is the key to measurement accuracy. Usually, we must determine that M_1 is really driven, and then write down the position coordinates of M_1. When recording the position changes of M_1, we should rotate the worm wheel codirectionally, and turning in reverse is forbidden during continuous measurements.

The surfaces of the optical components of Michelson interferometer, such as M_1, M_2, G_1, G_2, etc. mustn't be touched by fingers and wiped casually. The positions of G_1 and G_2 are crucial to the performance of the device, so they cannot be changed. Meanwhile, the worm gear system is very delicate, so when rotating the drum, avoid damaging the worm gear system, be slow and gentle, and don't be fussy or careless.

The position of reflector M_1 is read by the main calibrated scale near the left side of the guide rail (accurate to 1 mm), reading window 9 (accurate to 1/100 mm), and fine adjustment drum 13 (0.01 mm is divided into 100 small cells, means the precision is 10^{-4} mm), respectively, and it can be estimated to read to 10^{-5} mm. In practical measurement, read 2 numbers from the main calibrated scale and reading window, respectively; read three numbers from fine adjustment drum (including one estimate number); and then obtain the measurement data of seven numbers. The reading is shown as in Fig. 13 – 12:

(a)

(b)

(c)

Final result: 33.522,46 mm

Fig. 13 – 12 The reading of michelson interferometer
(a) Main calibrated scale (coarse adjustment handwheel);
(b) Reading window; (c) Fine adjustment drum

5. Experiment content and operation key points

(1) Adjustment of Michelson Interferometer

Michelson Interferometer is mainly used for observing various interference phenomena in physical optics, such as equal thickness interference, equal inclination interference, non-localized interference, localized interference and other forms of interference phenomena. The aim of the adjustment of Michelson Interferometer is to utilize it to obtain various interference phenomena;

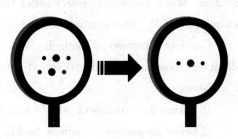

Fig. 13 – 13 The change of light spots in the frosted screen

therefore, Michelson interferometer should be adjusted to its prime state, i. e. M_1 and M_2 are strictly perpendicular to each other (adjust them by yourself), G_1 and G_2 are strictly parallel to each other (no need of adjusting them), and the angle between M_1 and M_2 and G_1 and G_2 is strictly 45° angle. In this part, one method that is easy to operate is introduced, i. e. adjusting it with He-Ne laser (see Fig. 13 – 5).

Lighten He-Ne laser to make the laser beam firstly get through a keyhole aperture and project the beam splitter to be divided into light ① and ② to irradiate M_1 and M_2, respectively. After reflected by M_1 and M_2, light ① will irradiate G_1 again. Part of light ① will reach the frosted screen used to observe through G_1. The other part of light ① is reflected at the front and the back surface of G_1, then returns to the back surface of the keyhole aperture, and forms a row of laterally aligned light spots. Adjust the three screws behind M_1, make the "brightest" light spot exactly coincide with the keyhole of aperture; at this time, we can also see the two brightest light spots coincide and interfere with each other in the frosted screen (when adjusting the instrument, the phenomenon changes of the frosted screen is shown in Fig. 13 – 13). Move away the keyhole aperture, set a small focal length convex lens coaxially between laser and interferometer to make the laser light beam converged to a point light source, make the point light source lighten G_1, and there will be the interference fringes on the screen. Adjust the two tension spring screws of M_2 slowly, move the center annular interference pattern to the center of the screen, and get the typical mode of the non-localized interference.

(2) Observe the phenomena of non-localized interference and measure the wavelength of He-Ne laser.

Adjust Michelson Interferometer according to the method mentioned above. When non-localized interference phenomenon appears, rotate the fine adjustment drum in one direction. Be patient until the center of the ring "bursts out" (or "disappears"), and notice the direction of rotation. When measuring, select better visibility as the standard, write down the prime position coordinates of M_1, and continue to rotate the fine adjustment drum in the same direction.

Concrete experimental methods: Whenever 50 rings "burst out" (or "disappear"), write down one coordinate; obtain 10 sets of data after 450 continual changing rings.

Notice:

①During the experiment, don't look at the laser beam without the beam expander;

②Before measuring, eliminate backlash (stripes appear to move inwards or outwards) and then write down the readings;

③When measuring, fine adjustment drum can only be rotated from one direction and changing the direction is forbidden;

④Don't over-tighten or over-loosen the screws attached to M_1 and M_2, and try to keep the tension of the spring behind the reflecting mirror.

(3) Observe localized interference phenomenon

①Observe equal inclination interference

Put the frosted glass between the lens and beam splitter G_1 to make previously scattering laser through frosted glass become plane extended light; use a telescope or eyes (at this time the eyes should focus at the infinity) to observe M_1 through G_1, and the circular interference fringe will be seen. Further adjust the two tension spring screws under M_1 to make it a perfect localized interference fringe.

②Optional experiment: observe equal thickness interference phenomenon

On the basis of equal inclination interference, rotate coarse adjustment handwheel to move the M_1 to make M_1 and M_2' approximate coincidence. At this time, the interference fringe becomes thicker, and finally almost disappears. Then adjust the two tension spring screws under M_2 to make a small angle between M_1 and M_2', and then equal thickness interference fringe will appear in the field of view, and the interference fringe spacing is inversely proportional to the angle φ.

Adjust the two tension spring screws under M_2 slightly to make the interference fringe thicker; rotate coarse adjustment handwheel to move M_1, and the bending-to-straight-to-bending changes of the interference fringe can be seen. At this time, change the light source with incandescent plane light, rotate the fine adjustment drum in the opposite direction to move M_1 slightly, and near the position, the colored white light interference fringe can be seen, i. e. there are some black and white straight fringes in the center, and colored interference fringes which are slightly bending.

Attention: fine adjustment drum must be used to get colored fringe; figure out the drum rotation direction that makes the fringes from slightly bending to straight. Don't be too hasty, or the more haste, the less speed.

6. Data recording and processing

According to the data in Table 13 – 1, process data with method of successive minus, calculate the wavelength of He-Ne laser wave and relative uncertainty, and obtain correct experiment results.

$$\begin{cases} \lambda = \overline{\lambda} \pm \Delta\lambda \\ E_r = (\overline{\lambda} - \lambda_{standard})/\lambda_{standard} \times 100\% \end{cases}$$

Table 13 – 1 Data recording table ($\lambda_{standard} = 6.328 \times 10^{-7}$ m)

$\Delta d_{仪} = 0.00005$ mm unit: mm

		d_0	d_{50}	d_{100}	d_{150}	d_{200}
	d_m					
		d_{250}	d_{300}	d_{350}	d_{400}	d_{450}
d_{m+250}						
Δd_{250}						
$\overline{\Delta d_{250}}$						
$\overline{\lambda} = 2 \times \overline{\Delta d_{250}}/250$						

7. Analysis and questions

(1) What are the main components of Michelson interferometer, what are the functions of them?

(2) When adjusting Michelson interferometer, why are there two lines of bright spots instead of two bright spots, how are these two bright spots formed?

(3) In what case does the interference fringe move "inwards" or "outwards", and in what

case is it sparse or dense?

(4) What is the reason of backlash, what impact does backlash have on measurement, how do we avoid backlash?

(5) Why cannot the rotation direction of fine adjustment drum be changed during the process of measurement?

Experiment 14

Franck – Hertz Experiment

1. Background and application

In 1913, the Danish physicist Niels Bohr (N. Bohr) in Rutherford, on the basis of combination of Planck quantum theory, successfully explains the stability of atoms and atomic line spectrum theory, and thus won the 1922 Nobel Prize for physics. According to Rutherford's atomic model, in the second year after the Bohr atomic theory was put forward, namely in 1914, Frank (James Frank, see Fig. 14 – 1) and Hertz (Gustav Hertz, see Fig. 14 – 2) with the method of the experiment proved the existence of the internal quantization level atom. They also proved that the atomic transition occurs absorption and emission energy is completely certain discontinuous. They completed the famous Frank-Hertz experiment, which provided direct and independently of the spectrum experimental evidence to Bohr Theory. The experiment played an important role in the development of the atomic theory, which become an important milestone in the history of the physics development. Because of the great scientific achievements, Frank and Hertz won the 1925 Nobel Prize in physics.

People have started electronic and atomic and molecular collision theory very early such as the gas discharge theory. Lenard (P. E. A. Lenard), in 1902, have already measured the ionization potentials of gas atom. But at that time, People didn't pay more attention to the changes that take place itself in the process of collision. They only cared about the atomic and electronic ionization due to collision.

At the beginning of the 20th century, the existence of atomic energy level was confirmed by studying atomic spectrum. Atomic spectrum of each line was formed because of the radiation by the atoms transition from a higher energy level. Frank improved Lenard Hertz's experiment method and used a very direct way to confirm the existence of atomic energy level. They used the method of collision between electron and rarefied gas atoms, observed the change of electron velocity before and after the collision. They found that when atoms collide, the electronic energy is always exchanged at a certain value. The experimental method was used to test the first excitation potential of mercury atom, and proved the existence of atomic internal quantization level.

Fig. 14 – 1　James Frank (1882—1964)　　Fig. 14 – 2　Gustav Hertz (1887—1975)

　　James Frank, German physicist, was born on August 26, 1882 in Hamburg and received his doctorate from the University of Berlin in 1906. In 1917, he worked as the director of the physical and chemical research institute of physics in William emperor. In 1921, he was employed by Georg-August-University working as a professor, and then immigrated to the United States in 1934. In 1935 and 1938, he worked as a professor in John Mr Perkins University and Chicago University successively. In 1955, because of the research in photosynthesis, he won the medal of the national academy of sciences. He was died on May 21, 1964, when accessing Georg-August-University. He is engaged in the atomic physics, nuclear, molecular spectroscopy and its applications in chemical and photosynthesis for his whole life. The main contribution he got in physics, is that, by electronic and atomic collision experiments, he earliest directly confirmed that Bohr about 1913 atoms stationary state hypothesis is correct. He also studied the electronic, atomic and molecular collision, atomic transition, energy levels of atoms, and the transfer of energy in the case of fluorescence in the system of atom, and illustrated the relationship between intermolecular forces and molecular spectroscopy. He put forward that the electron transition is quickly than molecules vibration, and then Frank-Condon principle was come out, which was as the basic principles of vibrational structure strength distribution of molecular electronic spectra.

　　Gustar Hertz, academy of sciences Berlin, Hamburg, was born on July 22. He was the discoverer of the electromagnetic wave, and nephew of H. Hertz. After graduated from Johnny's School in Hamburg, he came into the Georg-August-University in 1906, and worked as research assistant in Berlin university institute of physics in 1931. Because of the First World War, Hertz joined the army in 1914, and then back to Berlin in 1917 as the school teachers. In 1925 Hertz was selected as a professor and the director of Physics Institute at the Halle University. In 1928 he backed to Berlin, employed as the director of the teaching and research section in physics in Schloss Charlottenburg Industrial University. From 1945 to 1954, he was worked in the Soviet Union as the leader of a laboratory. During this time he was appointed as a professor and the director of Physics Institute in Leipzig University of Karl Marx. In 1975, he was died in Berlin. In his early time, he studied about the infrared absorption of carbon dioxide and the relationship between pressure and partial pressure. In 1913, he began to study electron collision with Frank. In 1928 Hertz back to Berlin, and his first task was to restore the physical institute, and responsible for the task of separating neon isotope using the method of multistage diffusion.

Frank-Hertz experiment is still one of the important means of exploring atomic structure. The method using rejects voltage to screen small energy electronic in experiment has become a widely used experimental technology. Through this experiment, we can know about his scientific method that used the macro process to reflect the micro process of collision between electrons and atomic, energy exchange and the state of energy change which difficult to directly observe. We can learn its ingenious ideas of scientific experiment; cultivate student's creative thinking and ability to solve practical problems.

2. Experiment principles

On the basis of Futherford's atomic model, Bohr atomic theory points out that:

(1) Electrons in an atom can move in some specific circular orbit, but not radiating electromagnetic wave, and the atoms is in a stable state, and has a certain energy.

(2) Atoms need to emission photons that atomic frequency v when they jump from the high-energy to low energy, and

$$h v = E_2 - E_1 \quad (14-1)$$

Where h is planck constant, $h = 6.63 \times 10^{-34}$ J·s.

The electrons of the atom in the normal state move in the first pathway, atomic energy minimum, and that is at the lowest level, this state is called the ground state. Atomic transitions from the ground state to a higher energy state called excited state. Energy required is called the critical energy when atomic transitions from the ground state to the first excited state. To change the state of atomic, the method is:

(1) Atoms itself absorb or emit electromagnetic radiation;

(2) Atoms collide with other particles and energy exchange. Franck – Hertz used the latter.

Franck-Hertz experiment apparatus principle shown in Fig. 14 – 3. In Franck-Hertz tube filled argon, electrons emitted by the thermal cathode, tend space $G_1 G_2$ under the action of $U_{G_1 K}$, and then accelerated movement to the grid electrode G_2 under the action of $U_{G_2 K}$. There is reverse voltage $U_{G_2 A}$ between plate electrode A and grid electrode $R_x = \dfrac{R_2}{R_1} R + \dfrac{r R_1'}{R_1' + R_2' + r}\left(\dfrac{R_2}{R_1} - \dfrac{R_2'}{R_1'}\right)$. As long as the electron energy is sufficient to overcome the rejection voltage $U_{G_2 A}$, electron can reach plate electrode A, the electron flow is formed. Micro-current electron flow meter can display the intensity。 Experiment maintained $U_{G_1 K}$, $U_{G_2 A}$ unchanged, $U_{G_2 K}$ increases, electron energy increases gradually, and also increases. In this process, electronics and argon atoms in tube elastic collisions, and electron energy can not be obtained by argon atoms. With the increase of $U_{G_2 K}$, when the electron energy is equal to or greater than the critical energy of argon atoms, all or most of the electron energy is passed to the argon atoms. The energy of the electrons decreases sharply, so that the electrons can not overcome the effects of rejection voltage, resulting in a sharp reduction of electrons that reaches plate electrode, then I_p decreased sharply. In this process, electronic and atom inelastic collision, argon atoms jump from the ground state to the excited state after getting energy from the electron. Continuing to increase $U_{G_2 K}$, electronic access to accelerate, and its energy increase, then electrons that overcome the rejection voltage and reached the plate electrode continue to increase, I_p increased. When the electron energy is equal to or greater than the critical energy of argon atoms, all or most of the energy of electrons passed to the argon atoms. Then electron energy reduced again, so that most of the electron can not reach plate electrode, it means that I_p decreased again. So, with $U_{G_2 K}$ continued to increase, the electron gets accelerated to reach plate electrode, I_p increases. As long as the electron energy is sufficient to overcome the rejection voltage $U_{G_2 A}$ again, argon atoms can obtain energy from the electron to achieve at excited state. Electrons lost all or most energy, so they can't reach plate electrode, and then I_p decreased sharply. With the increase of $U_{G_2 K}$, I_p changes in the performance characteristics of a distinct peaks and valleys. It shows in Fig. 14 – 4.

Franck–Hertz Experiment Experiment 14

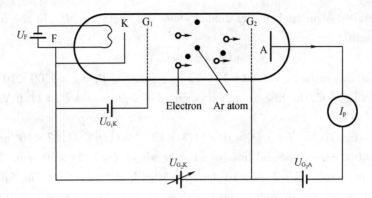

Fig. 14−3 Schematic diagram of F-H tube

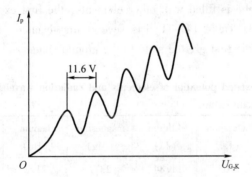

Fig. 14−4 I_p-U_{G_2K} relation map

This suggests that atomic absorption energy quantization characteristics. It found by curve $I_p - U_{G_2K}$: potential difference between two adjacent peaks are equal, and it is 11.6 V. When A is integer times of 11.6 V, I_p decreased sharply. When the electron energy is 11.6 eV, Argon atoms will absorb the energy of an electron and inspire. It can be considered that 11.6 eV is the needed energy of exciting argon atoms from the ground state to the first excited state. It can be seen, the potential difference between adjacent two peaks curve $I_p - U_{G_2K}$ is the first excitation potential of argon atom, and $U_0 = 11.6$ V. The critical energy value is obtained by argon atoms is:

$$eU_0 = E_2 - E_1 \qquad (14-2)$$

Atoms in the excited state is generally not a long time, it will automatically transition to the ground state and simultaneously release the energy values obtained. It is radiated in the form of light, its frequency is

$$\nu = eU/h \qquad (14-3)$$

Presumably, in the experiment radiation should be able to be seen that argon atoms transition from the first excited state back to the ground state. Its energy is eU_0, In the form of photons emitted wavelength

$$\lambda = hc/(eU_0) \qquad (14-4)$$

Where, planck constant $h = 6.63 \times 10^{-34}$ J·s; speed of light $c = 3.00 \times 10^8$ m/s; electron charge $e = 1.6 \times 10^{-19}$ C; The first excitation potential of argon atoms $U_0 = 11.6$ V. Combine with equation (14-4):

$$\lambda = 6.63 \times 10^{-34} \times 3.00 \times 10^8 /(1.6 \times 10^{-19} \times 11.6) = 1.07 \times 10^2 \text{ nm}$$

Franck-Hertz observed spectral lines of mercury whose $\lambda = 2.53 \times 10^2$ nm. This is in good agreement with the results, $\lambda = 2.54 \times 10^2$ nm calculated by $U_0 = 11.6$ V. This fully confirms the atomic level does exist. For exciting atoms to an excited state, atoms must absorb the energy of a certain magnitude, and the energy is not continuous.

If the Franck-Hertz tube is filled with other elements, the first excitation potential of these elements can be measured. Table 14 – 1 lists several first excited potential of elements and radiation wavelength from the first excited state to the ground state.

Table 14 – 1 Several first excited potential of elements and radiation wavelength from the first excited state to the ground state

Elements	sodium (Na)	potassium (K)	lithium (Li)	magnesium (Mg)	helium (He)	neon (Ne)	argon (Ar)
U_0/V	2.12	1.63	1.84	2.71	21.2	16.8	11.6
λ/nm	589.0 589.6	766.4 769.9	690.8	457.1	58.4	74.4	106.6

3. Experiment purposes

To understand, Frank-Hertz experiment design idea, principle and method; to measure argon atom first inspire potential and prove the existence of the atomic energy level with experimental method.

4. Experiment instruments

With the advent of the digital age, Franck-Hertz experiment have been developed from the analog era into the digital age. This generation of Fanke-Hertz experiment instrument is divided into manual and automatic operation, the parameters are all digital setting, and experimental data is a digital display. Computers display measurement interface, the same coordinate display multiple different experimental curve.

Intelligent Frank-Hertz experiment instrument is filled with argon, instrument panel is shown in Fig. 14-3, and is divided into eight zones by function.

①area, various input voltage connection socket and plate current output jack of Franck-Hertz tube.

②area, excitation voltage output jack required by Franck-Hertz tube. The left side is positive, the right side is negative.

③test current indication area. There are four current range gears, each range have a indicator light that indicates the currently selected current range block.

④test voltage indication area. Four voltage source select buttons are used to select different voltage source, each voltage source have a indicator light that indicates the currently selected current range block.

⑤test signal input output section. Current input socket insert Franck-Hertz tube plate current. Synchronization output and signal output jack can sent signal to an oscilloscope or computer display.

⑥adjust the keypad. It is used to change the current voltage source voltage setpoint and set check voltage.

⑦work status indication area. Communication indicator light indicates the communication state of experiment instrument and computer. Start key and work method key work together to complete the manual or automatic testing.

⑧power switch.

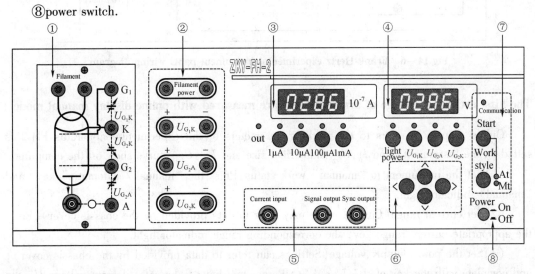

Fig. 14-5 Intelligent Frank-Hertz experiment instrument panel

5. Experiment content and operation key points

Preparation

(1) Be familiar with using the experimental apparatus of FH-2 method.

(2) Check the panel wiring of Frank-Hertz experimental instrument, as show in Fig. 14 – 6, confirming and start.

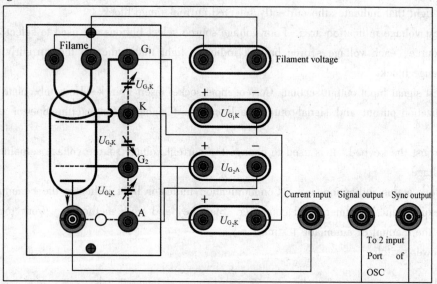

Fig 14 – 6　Frank-Hertz experimental instrument panel wiring diagram

The First excitation voltage of argon element are measured with online display manual mode

Online display is refers to the working parameters of the experimental apparatus FH-2 is setted on the instrument, and I_p, U_{G_2K} numerical size and $I_a - U_{G_2K}$ curve show on the computer.

(1) Set the instrument to "manual" work status, Press the "manual / automatic" key, and "manual" indicator light.

(2) Set current range (Current range may refer to data provided by the chassis cover), Press the appropriate current range key, the corresponding range indicator light.

(3) Set the power supply voltage (Settings can refer to data provided by the chassis cover), and complete with the key of ↓, ↑, ←, →. Parameters have to be set: filament voltage U_F、the first acceleration voltage U_{G_1K}、rejection voltage U_{G_2A}.

(4) The software works on "online display mode" is set to build between a computer and the FH-2 instrument.

(5) Press the "Start" button to start the experiment. Adjusting the voltage U_{G_2K} with ↓, ↑, ←, →. Start from 0.0 V. Step length is 1 V or 0.5 V.

(6) Record 6 peak data.

(7) Repeat steps (1) to (6), measured three times.

The first excitation voltage of argon element are measured with on-line testing automatic mode

On-line testing refers to the working parameters of the experimental apparatus of FH-2 is setted in computer software, transfer to FH-2 through RS232, the numerical size I_p, U_{G_2K} and $I_p - U_{G_2K}$ curve is displayed on the computer.

(1) Set the computer software work patterns for on-line testing, Set working parameters of the instrument through software.

(2) Set up communication link between a computer and FH-2 equipment. When the communication is normal, FH-2 instrument is in automatic mode, the voltage value is automatically increased by the set parameters.

(3) Print results.

Precautions

(1) Before open the power supply, please check the attachment, make sure it is connected correctly, if not, please report the teacher.

(2) Frank-Hertz tube is easily damaged due to improper voltage setting. Please set paramete refer to data provided by the chassis cover.

(3) In order to guarantee the uniqueness of the experimental data, the voltage U_{G_2K} must be one-way from small to large, the process can not be repeated. After recording the final data, the voltage U_{G_2K} rapidly return to zero immediately.

6. Data recording and processing

(1) Deal with experiment data of online display mode by the method of successive minus and obtain the average value $\overline{U_0}$. Comparative experimental value $\overline{U_0}$ and potential of subatomic the first excited state $U_0 = 11.6$ V, calculate the percentage difference, and write the resulting expression.

(2) Print online test-graph $I_p - U_{G_2K}$ (see Table 14 – 2)

Table 14-2 Date record form

unit: V

Peak(through) values	U_1	U_2	U_3	U_4	U_5	U_6
First set						
Second set						
Third set						
$\overline{U}$						

7. Analysis and questions

(1) By observing the experiment, we found that the plate current is not zero, what is the reason?

(2) From the visible-curve $I_p - U_{G_2K}$ plate current is not a sudden change, but has a smooth transition on every peak and valley, what is the reason?

(3) Can we use hydrogen instead of argon gas, why?

(4) In the experiment, How to cause the deviation between the first peak voltage and the first excitation voltage?

(5) Why do we add a Reverse rejects voltage between cathode and gate?

Photoelectric Effect Experiment

1. Background and application

In 1887, Hertz discovered that electrodes illuminated with ultraviolet light create electric sparks more easily, and this physical phenomenon is called photoelectric effect. After 1888, Hallwachs, Stoletov, Lenard, etc. studied this photoelectric effect for a long time and summarized photoelectric effect phenomena. However, these phenomena couldn't be explained in terms of classical theory. In 1905, Einstein boldly put forward the concept of "photon" in terms of Planck's blackbody radiation quantum hypothesis, successfully explained the photoelectric effect and set up the famous Einstein photoelectric equation that makes a forward step in understanding the nature of light and promotes the development of quantum theory. Thereafter, Milligan immediately made a detailed experiment research on photoelectric effect to prove the correctness of Einstein photoelectric equation and precisely measured Planck constant. As a result of Einstein and Miligan's great contributions in photoelectric effect, they were awarded the Nobel Price in Physics in 1921 and 1923, respectively.

Planck constant is related with the ubiquitous wave-particle duality and quantized energy exchange phenomenon in the microscopic world, and plays an important role in modern physics. Measuring Planck constant through photoelectric effect experiment can help students understand the quantum of light and universal constant h.

Light controlled apparatus made of photoelectric tubes can be used in automatic control, such as automatic counting, automatic alarm, automatic tracking, etc. Fig. 15 – 1 is the scheme of photoswitch, the working principle of which is as follows: when light illuminates the photoelectric tube, the circuit of photoelectric tube produces photocurrent that can be amplified by an amplifier to magnetize an electromagnet M to attract a magnet armature N; when there is no light on the phototube, there is no photocurrent in the phototube and electromagnet M will automatically release the magnet armature N. The rotating speed of some bodies can be measured by utilizing the photoelectric effect.

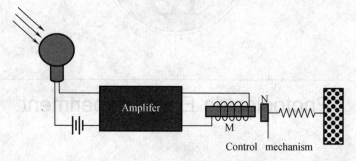

Fig. 15 – 1 The principle of photoswitch

Furthermore, various optoelectronic devices can be made with the photoelectric effect, such as photomultiplier, television camera tube, photoelectric tube, photoelectric meter. In this part, photomultiplier will be introduced, which can be used to measure very faint light. Fig. 15 – 2 is the structure of a photomultiplier: in the device, besides a cathode K and an anode A, there are several dynodes K_1, K_2, K_3, K_4, K_5, etc. Voltage should not only be applied between the cathode and the anode, but also between the dynodes, which makes the electric potential of the cathode the lowest, the electric potential of the dynodes gradually increase, and the electric potential of the anode the highest.

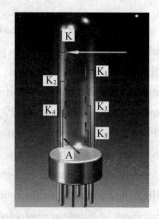

Fig. 15 – 2 The structure of a photomultiplier

Thus, accelerating electric field exists between two electrodes, and when the cathode is irradiated by light, it emits photoelectrons. Accelerated by the accelerating electric field, the photoelectrons impact the first dynode with large kinetic energy, and photoelectrons excite more electrons from the dynode. Accelerated by the accelerating electric field, the excited electrons impact the second dynode and excite more electrons. Thus, the number of excited electrons increases more and more and finally the number of electrons collected by the anode will be many times (usually $10^5 \sim 10^8$ times) of the number of electrons emitted by the cathode. Therefore, this device can produce very large current as long as it is irradiated by very faint light, and thus plays an important role in astronomy, military, etc.

Photoelectric effect also has an influence on sound film technology. Early films were silent. Although there were sounds in later films, the sounds were broadcast by panotrope to cooperate with pictures. The two machines were difficult to synchronize with each other, and thus the performance was not good. After shooting of the film, there will be sound recording. When recording the sounds, the changes of sounds are transformed into the changes of optical signals through special equipment, and thus shoot the "image" of the sounds on the edge of cinefilms to form dark stripes with different width, and this is the audio track on the edge of the film. During the movie playing, the "picture of sound" is reverted into sound by utilizing the phototube. The method is that: Irradiate the audio track by a very narrow light beam with the same intensity in the motion picture projector, and the light beam passes through the black strip and irradiates the photodetector on the other side. Since different parts of the audio track have different widths, thus during the movement of the film, the intensity of light that passes the audio track changes; when the changing light irradiates the phototube, there is a variable current in the circuit; after amplifying the current, and the sound is reverted through the loudspeaker. Optical audio track system is easy to be added to the film and be securely preserved during the entire life time of the film. When the sound film was firstly issued, the cinema projecting 35 mm film used this optical system. In the 1970s, Dolby Company realized stereo with two optical audio tracks. The system supplies the functions of stereo reproduction, surround sound effect, Dolby noise reduction, etc.

2. Experiment principles

When light with a certain frequency irradiates the metal surface, there are electrons escaping from its surface, and this phenomenon is known as the photoelectric effect. Its basic experimental facts are:

(1) Photoelectron emission rate (photocurrent) is proportional to the light intensity [see Fig. 15-3 (a)(b)];

(2) There is a threshold frequency (or cutoff frequency) of the photoelectric effect. When the frequency of the incident light is below a certain threshold value v_0, regardless of the intensity of light, no photoelectron will generate [see Fig. 15-3 (c)];

(3) The initial kinetic energy of photoelectron is independent of light intensity, but proportional to the frequency of the incident light [see Fig. 15-3 (d)];

(4) Photoelectric effect is an instantaneous effect, i. e. once light irradiates the metal

surface, photoelectrons are immediately produced. However, the experimental facts above can not be completely explained with Maxwell's classical electromagnetic theory.

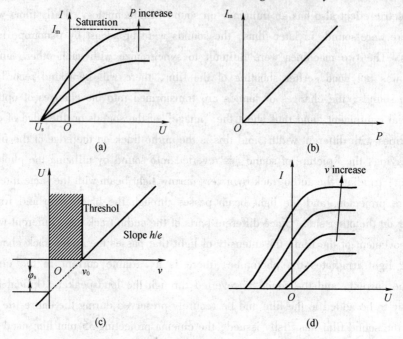

Fig. 15 – 3 Several characteristics of the photoelectric effect

Einstein believed that the light emitted from a point doesn't follow Maxwell electromagnetic theory, i.e. not in the form of a continuous distribution of the energy spreading into the space, but the light, whose frequency is v, radiates outwards one by one with the energy unit of hv. According to this theory, in the photoelectric effect, when a free electron in the metal absorbs the energy of hv from a photon of the incident light, if not losing energy due to collision on the way, a part is used for the work function W_s, and the rest corresponds to the largest kinetic energy after the electron escaping from the metal surface, i.e.

$$\frac{1}{2}mv_{max}^2 = hv - W \qquad (15-1)$$

This is the famous Einstein photoelectric equation. In this formula, h is Planck's constant, the known value is $6.626,075,5 \times 10^{-34}$ J · s. The formula (15 – 1) successfully explains the law of the photoelectric effect:

(1) When photon energy $hv < W_s$, the photoelectric effect can not be produced;

(2) Only when the frequency is greater than the threshold frequency of the incident light $v_0 =$

W_s/h, the photoelectric effect can be produced. The higher the frequency of the incident light is, the larger the initial kinetic energy of the escaping photoelectron is;

(3) The size of the light intensity means the size of the photon flux density, i. e. light intensity only affects the size of photocurrent. The size of saturated photocurrent is proportional to the size of light intensity of the incident light.

Fig. 15 - 4 is the experimental device of this experiment: the light, whose frequency is v and light intensity is I, irradiates on the cathode K of phototube, photoelectrons emitted from K move to the anode A, and form photocurrent along the outer loop. There is reverse voltage U_{KA} between the cathode and the anode, establishing deceleration field between K and A to prevent electrons from moving to the anode, so this method is also

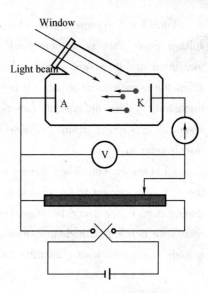

Fig. 15 - 4 The schematic diagram of photoelectric effect

known as deceleration field method. As the deceleration voltage U_{KA} (absolute value) increases, photoelectrons that reach the anodic will gradually decrease. Until photoelectrons with maximum kinetic energy are also blocked; photocurrent along the outer loop decreases to zero; at this time, initial kinetic energy of photoelectrons is all used to overcome the deceleration field, the relationship between the initial kinetic energy and the deceleration field satisfies this formula:

$$eU_s = \frac{1}{2}mv_{max}^2 \quad (15-2)$$

At this time, deceleration voltage U_{KA} is called cut-off voltage U_s. The cut-off voltage changes as the frequency of the incident light changes, i. e. the higher the frequency of the incident light is, the greater the cut-off voltage (absolute value) is. Make formula (15 - 2) into (15 - 1), and we get:

$$eU_s = hv - W_s \quad (15-3)$$

Because the work function of metal is an inherent property of metal, it has nothing to do with the frequency of the incident light. According to the formula (15 - 3), for the same kind of cathode, the relationship between cut-off voltage U_s and the frequency of the incident light v is linear, and the slope of the line is h/e. Thus, as long as measuring cut-off voltage U_s of the different frequencies of light, making a curve of $U_s \sim v$, and finding the slope of this curve, you can find the value of Planck constant h. Electron charge is $e = 1.60 \times 10^{-19}$ C.

Fig. 15 – 5 represents the change of photocurrent-voltage curve that is a theoretical case. In practical measurement, there are some unfavorable factors that affect the measurement results. If not making reasonable treatments to these unfavorable factors, it will bring great errors to experiment results. These unfavorable factors mainly refer to:

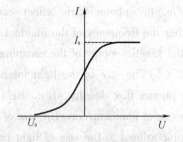

Fig. 15 – 5 **The volt-ampere characteristic of phototube**

(1) Dark current. When there is no light in phototube, there is weak current in the phototube under the effect of deceleration voltage, and because there is thermion emission at normal temperature and insulation resistance between cathode and anode is not enough high, etc. The relationship between the volt-ampere characteristic of phototube dark current and external voltage is basically linear.

(2) Anode emission current. The anode of phototube is made of high work function materials such as platinum, tungsten, etc. Because of depositing cathode materials, it will emit photoelectrons when meeting visible light. The electric field which has deceleration effect for electrons emitting from cathode is accelerating electric field for electrons emitting from anode, and it will make reverse saturation current in phototube. Avoiding the beam pointing at the anode directly is required when using the device, but the scattering light is inevitable from the cathode; therefore, reverse saturation current exists.

(3) The cathode of phototube is made of low work function materials. This material is still easy to be oxidized in high vacuum, so the work function of the cathode surface varies. With the increase of the reverse voltage, the photocurrent is not cut-off abruptly, but decreasing faster and gently reaching zero points; therefore, a high sensitivity galvanometer is needed.

Due to the above reasons, the $I – U$ relationship curve of phototube is shown in Fig. 15 – 6. Actually each current value on the measured curve includes three parts, i.e. two curves mentioned above and forward current produced by cathode photoelectric effect, so the voltage-current curve is not tangent to the U axis. Because the value of dark current is small compared with cathode forward current, their effects on the cut-off voltage can be ignored. The current emitting from anode is significant in practice, but it is subject to a certain rule. Through the analysis of these unfavorable factors, the reasonable design of phototube structure and the use of rational data processing method can reduce or even eliminate these disturbances. Two methods are usually adopted when determining the value of cut-off voltage:

(1) Intersection point method. The anode of phototube is made of high work function materials, so prevent possible evaporation of the cathode materials in producing process. Energize

the anode of phototube before experiment, reduce its sputtered cathode materials and avoid incident light directly irradiating the anode. In this way, reverse current can greatly be reduced, and its volt – ampere characteristic curve is very close to Fig. 15 – 5; therefore, the potential difference at the intersection point between the measured curve and the U-axis is approximately equal to the cut-off voltage U_s, which is known as the intersection point method.

(2) Inflection point method. Photocurrent emitting from phototube anode is comparatively great, but in structural design, if anode current can saturate as quickly as possible, volt-ampere characteristic curve will have an obvious inflection point after the saturation of the cathode current. As shown in Fig. 15 – 6, the potential difference at this inflection point is the cut-off voltage U_s.

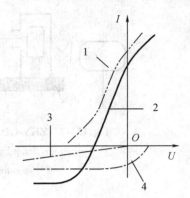

Fig. 15 – 6   The $I - U$ characteristic curve of phototube
1—ideal cathode emission current;
2—measured curve; 3—dark current;
4—anode emission current

3. Experiment purposes

Photoelectric effect is an important discovery of modern physics: It opens the door to quantum mechanics, and provides theoretical and experimental basis for the follow-up series of important discoveries. When previewing this experiment, briefly understand the discovery history of photoelectric effect; be familiar with the conditions of the photoelectric effect. During the experiment, deeply understand the basic laws of photoelectric effect and the quantum nature of light; have a solid foundation for the verification of Einstein photoelectric equation and the measurement of Planck's constant.

4. Experiment instruments

ZKY – GD – 4 photoelectric effect experiment instrument consists of a mercury lamp light source, color filter, aperture, phototube, tester (including phototube light source and weak current amplifier), the structure of which is shown in Fig. 15 – 7 and the master panel of which is shown as in Fig. 15 – 8.

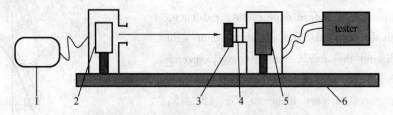

Fig. 15-7 ZKY-GD-4 photoelectric effect experiment instrument

1—mercury lamp light source;2—mercury lamp;3—color filter;
4—aperture;5—phototube;6—instrument base

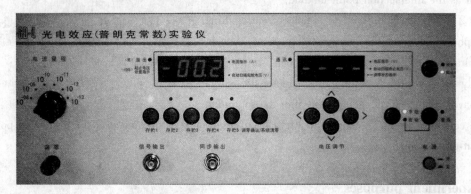

Fig. 15-8 ZKY-GD-4 photoelectric effect experimental instrument master panel

(1) Mercury lamp light source: The spectrum of the light source is in the range of 302 ~ 872 nm including 365.0 nm, 404.7 nm, 435.8 nm, 546.1 nm, 577.0 nm, 579.07 nm, etc., available for the experimental use.

(2) Filter group: There are five kinds of color filters, the central wavelengths of which are 365.0 nm, 404.7 nm, 435.8 nm, 546.0 nm, 578.0 nm, respectively.

(3) Aperture: There are three kinds of apertures, i.e. 2 mm, 4 mm and 8 mm.

(4) Phototube: Anode is a nickel ring, spectral ranges from 340 nm to 700 nm, and dark current is about 10^{-12} A.

(5) Phototube working power supply: Two voltage adjustment ranges are $-2 \sim 0$ V and $-1 \sim +50$ V, the three bits and half digital display, and stability $\leqslant 1\%$.

(6) Weak current amplifier: Current measurement ranges from 10^{-8} A to 10^{-13} A with six positions and the three bits and half digital display.

5. Experiment content and operation key points

The preparation work before experiment

(1) Cover the dark box baffle of phototube and port of mercury lamp.

(2) Turn on the tester and mercury lamp and preheat them for 20 minutes.

(3) Adjust the distance between phototube and mercury lamp (about 40 cm) and remain it unchanged on the instrument base, and make the port of mercury lamp point at the input port of the dark box of phototube.

(4) Connect voltage input terminal of phototube and voltage output terminal at the tester back panel with special wires (attention: red to red, and blue to blue).

(5) Rotate the selector button of current range on the tester panel to desirable gear (10^{-13} A). After fully preheating, adjust current to zero before test, that is, rotate "current zero" switch and the current indication will be 000.0. After the adjustment, press "zero adjustment confirmation/ system reset" switch, and the system comes into the test status.

The measurement of cut-off voltage

(1) Install aperture of $\Phi 4$ mm and filter of 365.0 nm at input port of phototube box (attention: mercury lamp cover can't be taken away during this process to avoid the damage of phototube).

(2) Set "volt-ampere characteristic test/ cut-off voltage test" switch at "cut-off voltage test", and the "current range" select switch at 10^{-13} A position.

(3) Set "manual/automatic" switch at manual mode, and open the lamp cover. Adjust voltage from low to high (absolute value reduces); observe the changes of current value; search U_s when current is zero; take its absolute value as the value of U_s of this wavelength, and record the data in table 15 – 1.

Table 15-1 $U_0 \sim v$ relation table

distance $L = 40$ cm   aperture $\Phi 4$ mm

Wavelength λ_i/nm	365.0	404.7	435.8	546.1	577.0
Frequency $v_i/(\times 10^{14}$ Hz$)$	8.214	7.408	6.879	5.490	5.196
cut-off voltage U_{si}/V					

(4) In turn, change the filters of 404.7 nm, 435.8 nm, 546.1 nm, 577.0 nm, respectively, and repeat the measuring step (3).

The measurement of volt-ampere characteristic of phototube

Use online test, and measure it with the computer running the main program to control the experimental instrument.

(1) Cover the dark box baffle of phototube and the port of mercury lamp. Adjust current range to 10^{-10} A, and adjust current to zero again; rotate "current zero" to make the current indication of 000.0; press "zero adjustment confirmation/system reset" switch, and make the system into the test ready state. Set working mode of volt-ampere characteristic test.

(2) Run an online testing program, enter the user's name and password as shown in Fig. 15-9, and a main program window will popup as shown in Fig. 15-10. Click the "start" button on the main window, and then a "new experiment" window popup as shown in Fig. 15-11; set the individual parameters in this window, such as the number of class, student's ID and name; select experiment item of volt-ampere characteristic, and click the "start" button; prompt dialog box of "whether current selection is correct" "whether the experimental instrument is adjusted to zero" "whether the experimental parameters are set correctly" as shown in Fig. 15-12 popup. If it is correct, click "Yes" and thus the "experimental parameter settings" window as shown in Fig. 15-13 will popup. Set the working mode of computer software, working parameters (class, student's ID, name, experiment item, working style, the number of curve, wavelength, frequency and the distance between phototube and mercury lamp, etc.). Dialog box of start-up experiment as shown in Fig. 15-14.

Photoelectric Effect Experiment Experiment 15

Fig. 15 – 9   System login interface

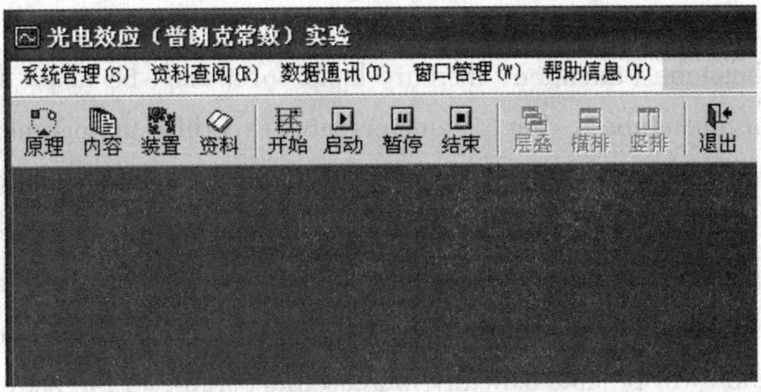

Fig. 15 – 10   Program main window

Fig. 15 – 11   New experiment window

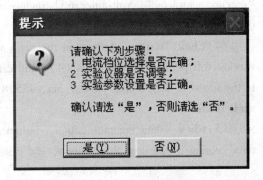

Fig. 15 – 12   Program dialog box of parameter setting

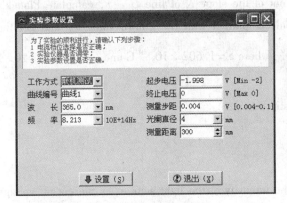

Fig. 15 – 13   Experiment parameter settings window

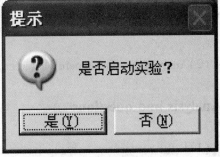

Fig. 15 – 14   Dialog box of start-up experiment

213

(3) Install aperture of $\Phi 2$ mm and the filter of 365.0 nm at the input port of dark box of the phototube (attention: mercury lamp cover can't be taken away in this process to avoid damaging the phototube). Set parameters at "experimental parameter settings" window, select "online test" as the working mode, select curves sequentially, select wavelengths according to the filter, and set the measuring distance as 400 mm. Click the "setting" button, prompted dialog box of "whether to start the experiment?" appears as shown in Fig. 15 – 11; take away the shade cover, and click "Yes" to start the experiment; after 30 seconds, the scanning voltage of phototube value will be automatically increased according to the setting parameters, and computer begins to draw volt-ampere characteristic curve. "The experimental curves" window below the status bar displays "communications"; after "experimental curves" window below the status bar displays "capture completed", volt-ampere characteristic curve corresponding to the wavelength of light is drawn.

(4) Click "startup" button on the main window, and the "experiment parameters setting" window will popup. Repeat steps (4), and draw volt-ampere characteristic curves corresponding to the filters of 365.0 nm, 404.7 nm, 435.8 nm, 546.1nm, 577.0 nm, respectively, in the same picture.

(5) Click "data communication" button on the main window, select "print experimental data" option in the drop down menu, and print the results.

6. Data recording and processing

(1) Use the regression method to find out of the slope K of line $U_s - v$, calculate Planck's constant according to the formula $h = eK$ and compare it with the known value h_0, and get the relative error by using the formula $E_r = \dfrac{h - h_0}{h_0}$, where $e = 1.602 \times 10^{-19}$ C and $h_0 = 6.626 \times 10^{-34}$ J $\cdot$ s.

(2) Analyze the volt-ampere characteristic curve.

7. Analysis and questions

(1) Is the photocurrent changing with the intensity of the light source, is the cut-off voltage changing with the light intensity?

(2) In theory, the intercept of line $U_s - v$ curve is the escape potential $\varphi_s(W_s/e)$ of cathode material. In practice, there is contact potential difference between the anode and cathode, so the

intercept of the measured curve is not equal to φ_s. Try to explain what produces contact potential difference, does it influence the experiment result?

(3) Discuss the importance of photoelectric effect on the setting of the quantum concept and the understanding of wave-particle duality of light.

Fundamental Experiment of Optical Fiber Sensors

1. Background and application

Optical fiber is a kind of special glass made by high refractive index glass within low refractive index glass casing, which made of preform drawing under the molten state, and then extrude to the double deck glass fiber with the diameter of micron. It can make the light transmission in its core. Compared with the usual waveguide, optical fiber is a wave transmission of high frequency electromagnetic wave. In the early 1770s, by observing that the light can transmit along the flowing wine stream, people knew that the light incoming at the end of the thin dielectric rod at a certain angle, due to that the medium index is different to air refractive index, and it made the light transmit forward in the rod between the air interface for total reflection. Deby and Hondros, in 1910, by using the theory of fluctuation analyzed the light path in dielectric. In 1927, Baird, Hansell pointed out that optical image can be transmitted by dielectric optical fibers. In the 50s, a kind of optical fiber structure with the high refractive center was invented. In 1950s optical fiber image transmission was applied to look inside the human body in medical treatment field. Transmission loss in fibers, at that time, was very huge, even in the most transparent optical glass, the loss was up to 1,000 dB/km. Obviously, it cannot meet the requirement of the communication. In 1966, Dr Charles Kao, the researcher of British standard telecommunication research institute, pointed out that the loss of light in glass is mainly caused by absorption loss of metal ions, and if the amount of these ions can be reduced to below 10^{-6}, the absorption loss of glass will be reduced under 20 dB/km. In 1970 a corporation in United States invented a high-purity silica fiber by using the theory of chemical vapor deposition, and its loss of

20 dB/km made it possible for long distance transmission. This achievement immediately drew attention from all over the world, and arose high enthusiasm of optical fiber communication research. The fiber loss can be reduced to 0.154 dB/km, which reaches the theory value of optical fiber loss limit.

With the development of the research on optical fiber, optical fiber is damaged by temperature, pressure and other influences of environmental, resulting in light intensity, phase, frequency and polarization optical parametric changes. Due to the optical fiber sensor has advantages such as high sensitivity, high pressure resistance, corrosion resistance, explosion proof, electromagnetic-interference resistance, wide frequency band, large dynamic range, soft, small volume, light weight and measuring without power supply, and make it be used in optical fiber sensor and the sensor array (see Fig. 16 − 1).

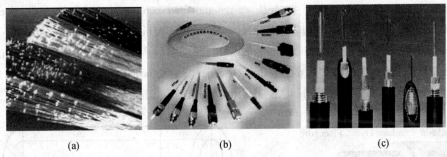

Fig. 16 − 1 The kind of optical fiber
(a) Optical fiber; (b) With optical fiber connector; (c) Fiber optic cable

Some advanced industrial nations have developed more than hundreds kinds of optical fiber sensors, which have been involved in military defense, aerospace, industrial agriculture, energy, environmental protection, biological medicine, metering, automatic control, and even daily life and other fields. The application of optical fiber sensor solves monitoring problem. Optical fiber not only initiated a revolution in communications technology, but brought vitality to sensing technology. Simultaneously, with the aid of optical fiber technology, a variety of emerging optical techniques and methods has also been developed, such as high power optical fiber transmission technology, optical fiber tweezers technology and moire interferometry etc. The application of photonic crystal fiber and double cladding fiber, plastic optical fiber, radiation-hardened infrared wavelengths in the optical fiber and other new special fiber will make important contributions to achieve highly information-based society.

The purpose of the developing experimental apparatus of optical fiber sensors is to be spread widely among the people. Not only does it enrich the teaching content, but also play an important role of extending the students' field of vision, the range of knowledge, improving learning interest.

2. Experiment principles

The principle of light in the optical fiber transmission

The light in the optical fiber transmission is based on optical of total reflection principle. Ordinary quartz silica (SiO_2) fiber structure is shown in FIg. 16-2 (a). Its structure includes fiber core and cladding and coating layer. Fiber core and cladding materials are silica, and the difference is that cladding material made of appropriate doping, mixed with a small amount of impurities (Ge, phosphorus, boron, fluoride, etc.), which makes the fiber core refractive index n_1 and cladding refractive index n_2. While $n_1 > n_2$, so when light is transmitted to the fiber core and cladding interface, when the incident angle is greater than the critical angle of incident light θ_C, light in the fiber core and cladding interface of total reflection occurs, as shown in Fig. 16-2 (b). It is because of this kind of total reflection thai limit the light transmission in the core of optical fiber.

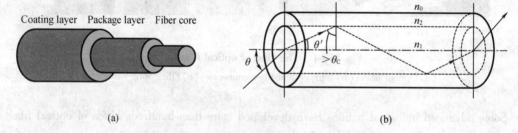

Figure 16-2 Optical fiber in optical fiber transmission schematic structure and light
(a) Optical fiber structure; (b) The light transmission

Fiber optic light field intensity distribution

Optical fiber emergent spatial distribution of the light field is as shown in Fig. 16-3 (a), in order to give a consistent with the actual, it is necessary to analyze the optical fiber at the end of the emergent light field. According to the mode theory of optical fiber transmission we could describe the light distribution as the gaussian distribution. Additionally, light approximately be as plane wave along the optical fiber transmission, and this plane wave emergence at the fiber end, which can be equivalent to vertical plane wave field incident to opaque screen hole on the surface.

The formation of circular aperture diffraction is shown in Fig. 16-3 (b). And the actual situation is close to a hybrid of the former situations. For analyzing conveniently, we make the following assumptions: ① the end surface of optical fiber: light field is made up of light intensity of uniform distribution along the radial plane wave along the radial direction which is gaussian distribution; ② emergent light field: fiber end emergent light field by quasi plane wave field of the circular aperture diffraction field and transmission of gaussian beams in free space superposition.

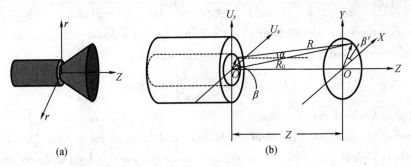

Fig. 16-3 Fiber end light field and the optical field grid analysis system
(a) Emergent light field; (b) The system coordinates

Under the above assumptions can be deduced theory formula (16-1) (please refer to the reference literature [1]):

$$I(r,z) = I_0 \left\{ p^2 \frac{a_0^2}{r^2} J_1^2\left(\frac{kr}{z}a_0\right) + q^2 \frac{(2\pi\omega_0^2)^2}{\lambda^2(4z^2 + k^2\omega_0^4)} \exp\left[-\frac{2k^2\omega_0^2 r^2}{4z^2 + (k\omega_0^2)^2}\right] \right\} \quad (16-1)$$

Equation, I_0 for light source coupled to launch the light intensity in optical fiber; p and q for two weight coefficient of the light field, meet the conditions, $p^2 + q^2 = 1$; a_0 is fiber core radius; $w_0 = \sigma \cdot a_0$ is gaussian beam radius; σ is the characterization of fiber refractive index distribution of relevant parameters; (r, z) is the receiver of the space coordinates; λ is Emergent light wavelength, k is wave number. Equation (16-1) shows that the fiber end emergent light field intensity distribution is gaussian distribution under different weights.

Fiber end is neither purely a gaussian beam, and beam of uniform distribution is not pure geometry. In order to better with the actual situation, we integrated the two approximation, and introduce the dimensionless blending parameter ξ and ξ as the light source, The numerical aperture of the fiber and light sources and optical fiber coupling of the composite modulation parameters help us get the following result:

$$\omega(z) = \sigma a_0 \left[1 + \xi\left(\frac{z}{a_0}\right)^{3/2} \mathrm{tg}\theta_C\right] \quad (16-2)$$

The process of actual use, the gradient refractive index fiber sometimes could be described $\sigma = \sqrt{\pi/2}$; For step refractive index distribution of optical fiber is usually described as $\sigma = \sqrt{\pi}$, for multimode optical fiber core diameter coarser, the diffraction effect basically is average, or described as $p \approx 0, q \approx 1$. Thus for large core multimode optical fiber, is easy to use equation (16-1) obtain the following form:

$$I(r,z) = \frac{I_0}{\pi\sigma^2 a_0^2 [1 + \xi(z/a_0)^{3/2} \text{tg}\theta_C]^2} \cdot \exp\left\{\frac{-r^2}{\sigma^2 a_0^2 [1 + \xi(z/a_0)^{3/2} \text{tg}\theta_C]^2}\right\} \quad (16-3)$$

If you send the same optical fiber in optical fiber end as a probe to the output light field in the receiver, the received light intensity can be expressed as

$$I_S(r,z) = \iint_S I(r,z) \mathrm{d}S = \iint_S \frac{I_0}{\pi\omega^2(z)} \cdot \exp\left[-\frac{r^2}{\omega^2(z)}\right] \mathrm{d}S \quad (16-4)$$

Here, S is received smooth surface.

Far area of the emergent light field on the fiber end, easy to meter, can be used to receive light intensity of optical fiber end face center as a whole the average light intensity on the surface of the fiber core, under this kind of approximation, we get in the receiving optical fiber terminals detected by light intensity formula

$$I(r,z) = \frac{SI_0}{\pi\omega^2(z)} \cdot \exp\left[-\frac{r^2}{\omega^2(z)}\right] \quad (16-5)$$

So the optical field distribution is shown in Fig. 16-4.

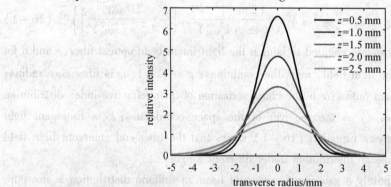

Fig. 16-4 Light field intensity fiber end points

Reflective optical fiber displacement sensor principle

Reflective optical fiber sensor principle is shown in Fig. 16-5. In which, A is composed of two optical fibers, optical fiber probe of a root to launch light, a reflection M, to accept a mirror plane mirror reflectance R. Will receive fiber and light optical fiber was issued by the mirror reflects light back to the field in image transform, equivalent to receive the light source directly into the light field of the fiber, the mirror, receive the light source and optical fiber spacing of $2z$, as shown in Fig. 16-5 (b).

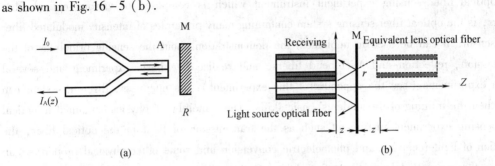

Fig. 16-5 double optical fiber reflection modulation principle diagram

(a) Optical fiber probe; (b) Equivalent light path diagram

Using the optical field distribution function image on fiber optic fiber end face of integral available light intensity is zero

$$I(r,2z) = \frac{S_A I_0 R}{\pi \omega^2(2z)} \cdot \exp\left\{-\frac{r^2}{\omega^2(2z)}\right\} \qquad (16-6)$$

Equation S_A is the receiving area of receiving optical fiber. Formula (16-6) which says the modulation function of reflective optical fiber probe.

3. Experiment purposes

Understand the basic structure of optical fiber and light in the optical fiber transmission principle. By means of the optical fiber end of emergent light field theory analysis and the quantitative measurement, to understand the fiber end emergent spatial intensity distribution of light field. It is a a kind of important scientific research approach about learning through the experimental data, sorting data, analysis data and looking for patterns and thinking about how to use them. To understand the working principle, application and calibration methods of the most simple optical fiber displacement sensor. To understand the working principle and application of intensity modulated optic fiber sensor.

4. Experiment instruments

Optical fiber sensing experiment instrument which is composed of various optical fiber sensors, is the optical fiber sensing system containing many principles of intensity modulated fiber optic sensor. It can be used to implement the demonstration about the sensing principles of the transmission, reflection and Microbend fibers, and realize five basic experiment and several design experiments (see the appendix of this experiment). Its openness, discrete system can strengthen the training of the students' basic skills. The principles of physics contained in optical fiber sensing experiment instrument such as the transmission of light in the optical fiber, the reflection of light receiving and photoelectric conversion and some other physical properties can enhance the perceptual knowledge of students. Optical fiber sensing experiment instrument has the advantages of simple structure, high sensitivity, good stability, convenient switching, wide application range etc.

The experiment was part of a series of experiments completed by optical fiber sensing experiment instrument. The experimental system is composed of the optical fiber sensing experiment instrument, LED light (green), optical fiber, PIN photoelectric detector (black), receiving optical fiber, a quasi-three dimensional micro displacement actuator, reflector, slightly curved modificator. Fiber optic sensing experimental instrument is shown in Fig. 16 – 6.

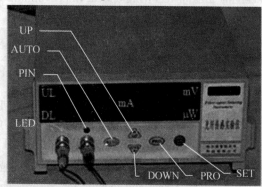

Fig. 16 – 6 Host experimental apparatus of optical fiber sensing

LED—light source output socket; PIN—light detector input socket;

AUTO—automatic step keys; PRO—programming control keys;

UP, DOWN—output current increasing or decreasing the key; SET—SET keys;

UL, DL, mA, μW—instrument displays a status indicator light

Fiber quasi three-dimensional displacement regulator with a light reflector and its three groups of fiber optic components are shown in Fig. 16 – 7.

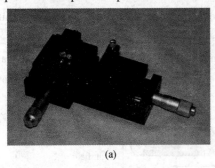
(a)

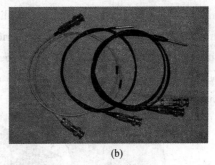
(b)

Fig. 16 –7   Form a complete set of experimental apparatus of optical fiber sensing component
(a) Quasi three-dimensional regulator; (b) Three groups of form a complete set of optical fiber

5. Experiment content and operation key points

In this experiment, three parts among five basic experiment of fiber optic sensing experiment instrument, namely the fiber end axial light intensity distribution measurement, the fiber end radial light field intensity distribution measurement [measuring device as shown in Fig. 16 – 8], and reflective optical fiber displacement sensor modulation characteristic curve measurement can be chosen.

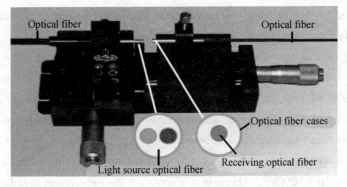

Figure 16 – 8   Fiber axial and radial light field intensity distribution measurement experiment device

The fiber end axial light intensity distribution measurement

In fact, light beaming into space from light fiber can irradiate a small area, but we want to measure the light power distribution of various points in the light irradiation area, when put receiving optical fiber in this area, it can be approximately regarded as point detection because the optical fiber core has a small area. The measurement of axial and radial light field intensity distribution is shown in Fig. 16 – 9.

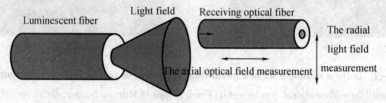

Figure 16 – 9 Light field optical fiber end measurement schematic diagram

Experimental operation steps of the axial optical field intensity distribution measurement

(1) Open the instrument power supply;

(2) Press to increase (decrease), adjust the working current to 5 ~ 16 mA (optical fiber end appears red light);

(3) Rotate spiral micrometer vertically, makes the optical fiber probe nearly as much as possible;

(4) Adjust the horizontal spiral micrometer and fiber optic height, make the output voltage the largest, at this time luminescent optic fiber core and receiving optical fiber core are aligned. counter-rotate the vertical spiral micrometer 1 ~ 2 small division (namely 10 ~ 20 μm, remove empty distance), record the current output voltage value.

(5) Continue to rotate spiral micrometer, record a voltage output values per 20 small division (namely 200 μm), end recording until output voltage value changes slowly (the fiber end axial optical field distribution measurement has been finished so far).

By checking the experimental data, measurement results should have similar regularity of the curve shown in Fig. 16 – 10 (a), otherwise it shows that the experiment failed.

The fiber end radial light field intensity distribution measurement

Experimental operation steps of the radial light field intensity distribution measurement

(1) Repeat steps (1) ~ (3);

(2) Rotate reversely vertical spiral micrometer 50 small divisions (namely 500 μm, or 0.5 mm), make the two optical fiber probe has a certain distance;

(3) Move one of the optical fiber probes horizontally until output voltage value of light intensity it has received changes slowly, continue to move 2 ~ 5 small divisions (The output voltage has no significant change at that time);

(4) Reverse move optical fiber probe 1 ~ 3 (eliminate idle), record the output voltage value;

(5) Continue to move the optical fiber probe along this direction, record output voltage value per five small divisions, until output voltage value of light intensity it has received changes slowly (the fiber end radial light field intensity distribution measurement has been completed so far).

By checking the experimental data, measurement results should have similar regularity of the curve shown in Fig. 16 – 10 (b), otherwise it shows that the experiment failed.

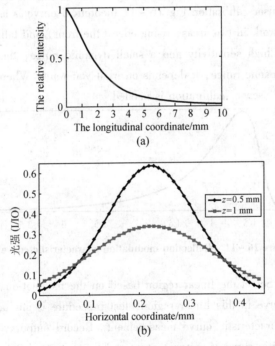

Figure 16 – 10 Light field intensity distribution of optical fiber end theoretical curve
(a) The optical fiber end axial distribution of light field intensity;
(b) The optical fiber end of radial intensity of optical field distribution

Reflective optical fiber displacement sensor modulation characteristic curve measurement

Take down the optical fiber in Fig. 16 – 8 from the fine-tuning frame, install the reflection receiving optical fiber in fine-tuning frame, and make the optical fiber probe alignment with reflector on the trimming frame. The experimental steps are as shown:

(1) Connect the power, make the LED drive current adjust to the specified current value (40 mA).

(2) Adjust the vertical micro adjustment knob, make the detection fiber move to the location contact with the mirror surface.

(3) Rotate micro adjustment knob along the longitudinal direction away from the mirror, record micrometer readings and the corresponding output voltage value per adjusting 0.1 mm (10 small divisions).

(4) Displacement sensor calibration Fig. 16 – 11 theoretical curve is as shown, optical fiber displacement sensor can work in two areas, rising edge (the front) and falling edge (the back). The rising edge area has high sensitivity and a small dynamic range, the falling edge has low sensitivity and a large dynamic range, it depends on what you want. When it is used as a fiber-optic displacement sensor, sensor calibration is needed.

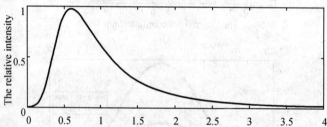

Figure 16 – 11　Reflection modulation characteristic curve

Calibration method: Select the linear region based on the modulation curve characteristics, then give a calibration curve in this linear region, test procedure is similar to the experimental content of modulation characteristic curve measurement. Record output voltage value every 50 μm, make the characteristic curve of voltage output and the distance between optical fiber probe and reflector. So when made measurement, the distance between the optical fiber probe and reflector is measured based on the modulation characteristic curve.

6. Data recording and processing

The fiber end axial light intensity distribution measurement

(1) Fiber end axial light intensity distribution measurement;

(2) Briefly describe the characteristics of the axial optical field distribution.

The fiber end radial light field intensity distribution measurement

(1) Trace point with coordinates paper or draw the experimental curves in Excel (data record form can be designed by yourself);

(2) Briefly describe the characteristics of the radial optical field distribution combined with the axial optical field distribution curve.

Reflective optical fiber displacement sensor modulation characteristic curve measurement

(1) Trace point with coordinates paper or draw the experimental curves in Excel (data record form can be designed by yourself);

(2) Give the linear polynomial fitting in chosen linear segment.

7. Analysis and questions

In principle, all physics which can be turned into a tiny displacement are measured by reflective optical fiber probe. By use of a calibrated reflective optical fiber sensing probe, think how to measure a physical quantity transmitted displacement such as the changes of length (displacement measurement), the changes of bimetallic strip with temperature (temperature measurement), changes of diaphragm with pressure (pressure measurement), etc. This experiment can be designed by yourself.

8. The application of experimental apparatus of optical fiber sensing of appendix (reference)

LED light source, I – P characteristic curve test

Commonly used semiconductor Light sources are LED (Light Emitting Diode) and LD (Laser

Diode). Fig. 16 – 7 (b) shows that the receiving optical fiber is directly connected to the experimental apparatus of optical fiber sensing. There is no need for fine-tuning, then we can complete the I – P characteristic curve of LED light source test experiment.

Background knowledge

In the field of optical fiber technology, light source mainly serving as light energy supply components, is one of the foundations of optical fiber technology. Optical fiber technology has some basic requirements to the light source. First of all, the peak wavelength of light source should be under the low loss window of optical fiber, requiring less material dispersion; Secondly, light output power must be large enough, the fine power shall generally be from 10 μW to several milliwatt; Thirdly, with high reliability, lifetime is at least 100,000 hours, which can meet the engineering requirement; Fourthly, the light source should be easily operated, and the modulation rate should be able to meet the requirements of the system; Fifthly, electro-optic conversion is quite efficient, otherwise it will cause severe fever and short longevity; Sixthly, light source should save electricity, and light volume, weight, should not be too huge. The vast majority of light sources are made of semiconductor materials in general, which can be divided into two kinds: one is called the LD of narrow linewidth coherent light source, another is called LED wide spectrum of incoherent light. Several common LED/LD are shown in Fig. 16 – 12.

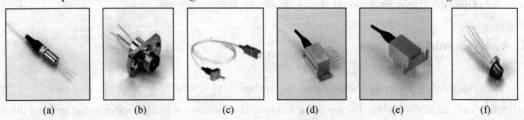

Figure 16 – 12 Kinds of LED/LD semiconductor light source
(a)1,310 nm LED;(b)1,310 nm SLD;(c)1,550 nm LED;
(d)1,550 nm DFB Lasers;(e)F – PLD;(f)MQW – DFB LD

Luminescence properties of P – N junction

Semiconductor light source is mainly composed of P – N semiconductor structure. In order to better describe P – N structure, we introduce some basic concepts of semiconductor physics below.

(1) The communization movement

Atoms in the semiconductor, like other crystal materials, are closely arranged according to certain rule. Due to the cohesive strength between the atomic interaction, individual atoms keep a certain distance, and the binding force is called covalent bond. Due to adjacent atoms in the semiconductor single crystal are so close, and composed single crystal materials not only be effected by the action of the atoms of electrons within atoms, but also be controlled by the effect of adjacent atoms. Adjacent atomic electron orbital (quantum state) will have a certain degree of overlap, and by overlapping orbits, electrons can be moved from one atom to another, which is called communization movement.

(2) The valence band and conduction band, the band gap

Because of the electron in the crystal communization movement can obtain different speed, corresponds to an atomic energy level, the level of quantum state is not a single, but is divided into N level of energy which is very close to each other (N for crystal atoms contained in the atomic number). Due to that the material N is large, the number of atoms in the gap between the level is very small, so a set of intensive level band is formed.

Inner electron energy level is fully filled with general atomic electronic, and their communization movement are weak, so the corresponding band is narrow. When atoms become crystal, the inner level corresponding to the band are also filled up with electronic, which is called a full band. The outermost layer of the valence electrons which is effected by atoms bound function are weaker, but the effect of adjacent atoms is strong. As a result, the communization movement is strong, so the corresponding band is wide. Valence electrons filled band is called the valence band and above the valence band is basically empty, known as empty band, which is called the conduction band. Band gap between the valence band and conduction band is called forbidden band and the forbidden band width represents a band to another band energy difference.

In the case of low temperature of absolute zero, the crystal of the electrons are in the valence band, and the conduction band is completely empty. According to the principle of Pauli, they cannot be conductive if electrons in the valence band are heated or light excitation, and the excited electron would have been a transition to the conduction band. So that we can produce electric current, which can be conductive to crystalline materials. We wrote the bottom of the conduction band energy for E_C; write on the top of the valence band energy for E_V. There is no electronics between the E_C and E_V. We consider the difference between the E_C and E_V as E_g, known as the forbidden band width. We consider the case that E_g is larger. For insulation material, because the forbidden band the width is very big, electrons in the valence band is very difficult to transit into the conduction band, so it shows a good electrical insulation. Conductor material E_g is small, so it shows good electrical conductivity. Semiconductor forbidden band width between conductor and insulator, thus its conductive ability is also between them. When the electrons in the valence band are inspired to the conduction band, and leave an electronic space in the valence band. The role of space as a positively charged particle, in semiconductor physics, which is called a hole. Absorb the energy of electrons in the valence band and transit into the conduction band, forming hole in the valence band. Similarly, in the conduction band, electrons transit to the valence band, and the valence band will fill the hole. The process is called composite.

P type semiconductor, N type semiconductor and the P – N junction

In the actual semiconductor, single crystal materials, often exist and the atoms that make up the crystal substrate different, and all kinds of defects appeared in the process of crystal formation. Material, the purpose of purification to remove harmful impurities, eliminate or reduce some defects. However, the impurity atoms plays an decisive role in semiconductor by adding trace amounts of impurities (hereinafter referred to as doping) to control. According to different, doped electrical semiconductor and hole type semiconductor materials can be obtained.

With impurity and defect very pure and complete semiconductor is called the intrinsic semiconductor. As shown in Fig. 16 – 13 (a). Its characteristic is, in the semiconductor material, the number of conducting electrons and valence band equal to the number of holes. Usually called intrinsic semiconductor type I semiconductor. So-called electrical semiconductor (N type semiconductor) is through the doping making the number of electrons in conduction band price comparison with the number of holes larger semiconductors, as shown in Fig. 16 – 13 (b). The so-called hole type semiconductor (P type semiconductor), is through the doping makes the number of the valence band hole is much greater than the number of conducting electrons semiconductors, as shown in Fig. 16 – 13 (c). Semiconductor physics theory analysis and experimental results show that the physical properties of semiconductor depends largely on the amount and type of impurities contained. Combine the different types of semiconductor, which can be made into all kinds of semiconductor devices.

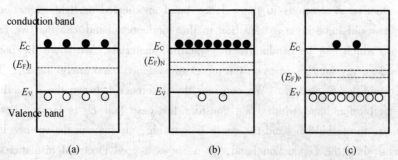

Fig. 16 – 13   N – P type semiconductor band diagram

(a) Intrinsic semiconductor; (b) N type semiconductor; (c) P type semiconductor

P type semiconductor and N – type semiconductor interface are called P – N junction, which are many semiconductor devices (including laser diode, light emitting diode) the composition of the core.

In P – type semiconductor with extra hole in the N – type semiconductor with redundant electronics, when combine these two kinds of semiconductor, the hole in the P spreading N area, near the interface left with negatively charged ions, and area of N electronic spread P area; Near the interface left the positively charged ions. In this way, on both sides of interface are formed with the opposite charge area, which is called the space charge region. Formed from the opposite charge, a self-built electric field, the direction is made up of N to P area, as shown in Fig. 16 – 14. Because of the existence of self-built electric fields on both sides of the interface creates a potential difference of V_D, the spread of the electric potential difference prevented the hole and electron, the final balance, as shown in Fig. 16 – 15. Therefore, we called the V_D block hole and electron diffusion barrier. According to theoretical analysis, it can use an energy E_F (called the Fermi level) to describe the distribution of the electrons and holes. For E_F level below, the possibility of electrons occupying more than 1/2, the possibility of a hole to occupy less than 1/2; For E_F level, is the opposite. Under the equilibrium state, P and N area have unified Fermi level. For P area, because there are many holes in the crystal, so the top of valence band is near the Fermi level. For N area, because of the crystals in many electronic, so the bottom of the conduction band is near the Fermi level.

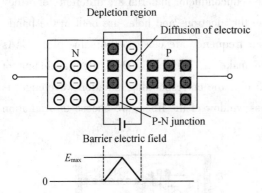

Figure 16 – 14 P – N junction barrier electric field

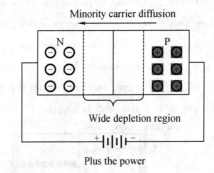

Figure 16 – 15 P – N junction in the nature of the equilibrium state

The luminescence of P – N junction

The luminescence mechanism of LED is the positive electro-optical properties of P – N junction. Adding a positive voltage V to the P – N junction leaded to that a part of the barrier voltage is cancelled out, and remained the voltage that $(V_D - V)$. The positive voltage added destroys the previous balance, as a result, the Fermi level between area P and N will be separated. Then we can describe the distribution of electron and hole by using two so-called expectant Fermi level. We denote the expectant Fermi level in area N as $(E_F)N$, for the level less than $(E_F)N$, the possibility that electrons occupy is more than $1/2$; and the one in area P as $(E_F)N$, for the level less than $(E_F)P$, the possibility that holes occupy is more than $1/2$. When we add a very large positive voltage to the P − N junction, the hole in area P will be mostly injected into area N, and the electron in area N will be mostly injected into area P, at the interface, the electrons and the holes will be getting together, then electrons will release energy as much as band gap E_g, in condition of radiative transition, the energy will become a photon which as a frequency of $v = E_g/h (6.625 \times 10^{-34} \text{ J} \cdot \text{s})$, where h is planck's constant, as is shown in Fig. 16 − 16.

LED is a lightening device which can be directly injected current, when adds a positive voltage to P − N junction of LED, the electrons and holes are getting together, which leads to radiate the lightwave at a frequecy of $v = E_g/h$, but it does not have a fixed phase relationship, it can have a different polarization direction, and each light beams spread to every possible direction, which is called self-radiation transition. The radiation wavelength λ can be expressed as $\lambda = hc/E_g = 1.2396/E_g$.

The values E_g among different Monocrystalline semiconductor material are different, and they have different lightening wavelengths, it is because the electrons and holes are both in the band, and they also have different energy levels, but their frequency are close to the value of v. GaAs has an E_g as 1.435 eV, so we can use GaAs to make an infrared LED at the wavelength of 0.85 μm, In GaAs P has an E from 0.75 to 1.35 eV, the corresponding radiation wavelength is from 1.65 to 0.92 μm, thinking about the low loss window of the fiber, we choose the radiation wavelength at 1.3 μm and 1.55 μm

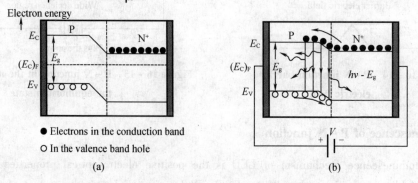

Figure 16 − 16 Semiconductor LED light emitting process

The spectrum from LED has a certain width, it is because the energy bands have certain width, as a result, the starting and ending of transition have a certain width; indeed, the compound in semiconductor is a complicated process, there are also transitions between conduction band and impurities, impurities and impurities, etc. This experience used an LED at the center wavelength of 650 nm.

The spectrum characteristics of LED semiconductor light source

The spectrum characteristics is a curve distribution between intensity (or energy) and wavelength (or frequency). The spectrum can be effected by material type and luminescent center structure, but not by the geometry and the package. We use peak wavelength and half line with to describe the spectrum.

In the image of lightening spectrum, we denote the wavelength where the value of intensity is largest as peak wavelength λ_P, and the wavelenth difference $\Delta\lambda$ between the half intensity point λ_1 and λ_2 as half line with, as is shown in Fig. 16-17. The width of the spectrum can also defined as the 1/e of intensity, or the wavelengths difference between two points which has a difference of 1/100 of wavelengths, please pay attention to the specific definition methods.

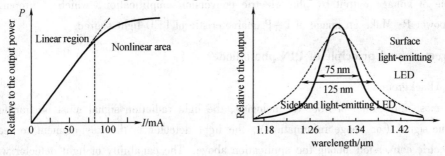

Fig. 16 - 17 Typical LED light source, I - P characteristic curve and spectrum diagram

The peak wavelength is usually related to the material, the common semiconductor light resources in fiber technology are at the wavelength of 0.85 μm, 0.98 μm, 1.31 μm, 1.48 μm and 1.55 μm. For the semiconductor LED light source, because there are not optical resonators to choose wavelength, the spectrum is often spontaneous radiation, which usually has a large width from 30 nm to 40 nm.

Spectral characteristics are usually being the basis to choose a light source. The factory would have been providing spectral image of every device before they came out from the production line.

The lifetime of the light source

Users also care about the lifetime of light sources. For the semiconductor LED, the lifetime is the time that the brightness turns from full to the half. The LED's lifetime, because of the stability, is usually very long from 106 hours to 109 hours. But that in semiconductor lasers will be very short. In the year 1970 the semiconductor lasers which have a good photoelectric conversion had been produced, however, people found that the photoelectric conversion of it has degenerated quickly, and they have a short lifetime. The lifetime of the semiconductor lasers producted before 1973 is below 1,000 hours, and through continuous research and improvement, the lifetime of the ones producted now is more than 100,000 hours, which can meet the needs of optical fiber communication.

Experimental contents and operating key points

Take out the fiber, gather the light source with the socket of LED, gather the detector with the socket of PIN detector; Turn the power on, the "LED current" window will indicate the current of LED(mA). Adjust the current adjusting button to make the current turn least; record the value of voltage output by photoelectric conversion amplification, which is proportional to output power P; Make an image of I – P characteristic of LED light source.

The structure and principle of PIN photodiode

(1) Background

In most application of fiber technologies, the light radiation signal must be transformed to electronic signal (or image information), the light detector is the key element to realize the photoelectric conversion among the application above. The capability of light detector will effect the capability of whole detecting system. On the other hand, use transforming light signal to electronic signal can realize the displaying or controlling function. The light detector can not only replace human eyes, but also be the elongation of human eyes for its wide spectral response. In the past decades, people have tried their best on light detection technology, in which the greatest achievement is that they produced the light detectors that have many types and functions based on the development of semiconductor physics and solid state physics, in this case they realized to invent the detector with high sensitivity and high integration. Under the development of fiber-communication, coherent light communication, fiber optic sensing technology and optical radar technology, the demand for the sensitivity of light radiation detector has been higher and higher, the detector and its detecting technology have been paid more attention.

The light detectors can detect the light radiation because the light radiation transports energy. Light radiation incidents to the detector make it produce photo-generated carriers or change its own characteristics such as temperature. According to the different respond modes between them, the former is called photoelectric effect, the latter is called photo thermal effect, so the detectors like above are called photon detectors and heat detectors. And the light detectors can be divided into photoelectron emission detectors, photoconductive detectors, photovoltaic detectors and photon traction detectors; the heat detectors can be divided into thermocouples, calorimeters, pyroelectric detectors and pneumatic detectors.

The species of detectors are large, and the working mechanisms and modes are different, but in the application of the detectors, we need that the spectral response can correspond to the three low loss windows, which are at 0.85 μm, 1.31 μm and 1.55 μm. And we need the detectors to have high conversion efficiency, a long lifetime and a stable work. After a long practice we know that the photovoltaic detector PIN semiconductor photodiode and optical avalanche diode can meet the above requirements. In addition, it has characteristics of small, convenient and compatible with electrical equipments, as a result, it has been widely used in optical fiber technology. Typical light detector is shown in Fig. 16 – 18.

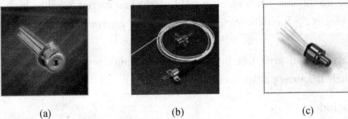

Fig. 16 – 18 **Optical fiber technology in several typical photodetector**
(a) Si – PIN photodiode; (b) GaAs – PIN photodiode; (c) PIN – TIA receiving component

(2) PIN: photodiode

The P – N junction, with the photoelectric effect, actually is not a complete mature detector. Because of the simple structure, can't reduce the dark current and improve the response rate, and the stability of the device is also very poor. PIN photodiode improved the structure of P – N junction, it has an intrinsic semiconductor between the P – type and N – type layer, forming the structure of PIN.

The typical application of PIN(or PIN – PD) is showed in Fig. 16 – 19. In this figure, V is reverse bias voltage; R_L is the load resistance. Light waves enter from surface P (it also can be lighted from surface N or layer I, and this should be decided by different design requirements). when the device is in the state of reverse bias, the power supply generates electric field E has the same direction of the built-in electric field E_i; integrating E and E_i we obtain the electric field $E_p = E_i + E$, and make the depletion region w significantly broaden. Additionally, the intrinsic layer has a very high resistance, close to the insulator, and it makes the depletion zone extend throughout the whole extrinsic region. The result brings three positive aspects:

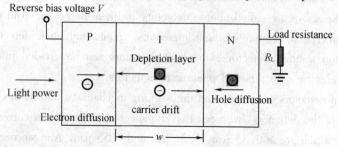

Fig. 16 – 19   PIN photoelectric diode reverse bias light detection

First, region I is thicker than that of P and N areas, the incident light energy inspires carrier in a wide range, increasing the chance of producing a carrier, and thus improve the response rate of the device.

Secondly, the whole area I has electric field, light raw carrier reaches high diffusion velocity faster than drift velocity, which increases the chance of generating carriers and it improves the response speed.

Thirdly, the exhausted area has been extended, which reduces junction capacitance, and conduces to the high frequency response.

Avalanche photodiode

Another kind of commonly used photoelectric detector is avalanche photodiode (APD). Increasing reverse bias of the photodiode results in the increase of electric field and drift speed inside the P – N junction. When electric field intensity increase to a certain value, the high speed carriers dislodge the lattice and makes it explode secondary electron, which will impact the original ones that leads to more electrons explodes, the phenomenon above is called impact ionization, which is a chain reaction. It leads the avalanche increasing of carriers, and the light current outside increases accordingly, this is the principle of APD, which has an internal current gain, the optical communication receiver made by APD has a higher sensitivity. But its noise is much larger than that in PIN, which limits the improve of the sensitivity.

Fiber optic microbend type pressure/displacement sensor

(1) Background knowledge

In the field of optical communication, the fiber bending loss of light intensity has drew lots of attention. , D. Marcuse and D. Gloge research on fiber bending and mode coupling which has great significance in the field of loss of bending optical fiber. With the development of optical fiber sensor technology, nowadays, the loss caused by microbend has become a useful sensor technology. By using the micro bending of the fiber can measure various physical parameters.

Optical fiber bending loss can be divided into two categories: the Macro bending loss and the Micro bending loss. Bending loss of optic fiber sensor is made by both bending macro and micro bending. The former includes light path bend, loose optical fiber cable etc. Its characteristics are long relative length, narrow spatial frequency distribution. The latter mainly causes by small deviation to the fiber's flat stat, and its characteristics are short relative length, wide space distribution. When the fiber was manufactured to cable in the low temperature circumstance, the winded process causes tension and deformation.

Fig. 16 - 20 shows the explanation to physical bending loss of optics fiber. In long and straight fiber phase surface is vertical to axis. But if the optical fiber bend is as shown in Fig. 16 - 20, the string with constant radius, field and wave front will revolve around the center of curvature of bending parts. So the phase velocity which is parallel to the optical axis must be linearly increased with the distance to the center of curvature C increased, due to that the uniform optical fiber cladding and phase velocity should not exceed the velocity of light at that place, so in curved surface there is a specific radius R, over which phase velocity will not increases, and the field also must become radioactive field, as shown in Fig. 16 - 20.

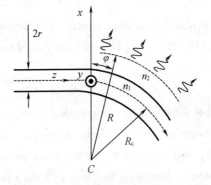

Fig. 16 - 20 Bending into a curved fiber radius

(2) The experimental principle

The principle of microbend fiber-optic sensor structure as shown in Fig. 16 - 21. When optical fiber bend, due to that its total reflection conditions are destroyed, some beams enter into the cladding and cause energy loss in fiber core. In order to expand this effect, we stick the fiber in a comb structure with cycle of wavelength Λ. When the comb structure (deformation) is stressed, the shape of fiber changes, and energy leaked into the cladding will change. Approximately, optical fiber could be considered as a sine curve model, and the curve function is

$$f(z) = \begin{cases} A\sin\omega \cdot z (0 \leqslant z \leqslant L) \\ 0 (z < 0, z > L) \end{cases} \qquad (16-7)$$

In which, A is the amplitude of comb structure; L is fiber optic microbend area; $\Lambda = 2\pi/\Lambda$ is the frequency of its bending; Λ is bending cycle length of comb structure.

Energy loss coefficient caused by optical fiber bended is

$$\alpha = \frac{A^2 L}{4}\left\{\frac{\sin\left[(\omega-\omega_c)\frac{L}{2}\right]}{(\omega-\omega_c)\frac{L}{2}} + \frac{\sin\left[(\omega+\omega_c)\frac{L}{2}\right]}{(\omega+\omega_c)\frac{L}{2}}\right\} \quad (16-8)$$

In which, ω_c is called resonance frequency, and it can be expressed as

$$\omega_c = \frac{2\pi}{\Lambda_c} = \beta - \beta' = \Delta\beta \quad (16-9)$$

In which, Λ_c is resonant wavelength; β and β' are two constants; when $\omega = \omega_c$, the two patterns of optical power coupling are tight, thus loss also increases. If we choose two adjacent modes, we can obtain

$$\Delta\beta = \frac{\sqrt{2\Delta}}{r} \quad (16-10)$$

In which, r is fiber radius, Δ is a relative different index between fiber core and cladding. From (16-9) and (16-10) we can obtain

$$\Lambda_c = \frac{2\pi r}{\sqrt{2\Delta}} \quad (16-11)$$

For the communication optical fiber, $r = 25~\mu m, \Delta \leq 0.01, \Lambda_c \cong 1.1~mm$.

Formula (16-8) shows a loss (α) and the square of bend is proportional to the length. Usually, we make optical fiber pass through a comb structure with circle Λ to generate the microbend. According to (16-11) the value of Λ is generally too small. In fact we could chose odd number, like 3, 5, 7, etc., as shown in Fig. 16-22.

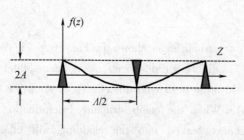

Fig. 16-21 Fiber optic microbend modulation principle diagram

Fig. 16-22 Microbend modulator

The experiment content and key points for operation

Embed the microbend deformer to three-dimensional micro displacement adjuster. With diameter of 50 μm, two ends of the measured optical fiber are respectively encapsulated with LED light and the PIN photodetector Q_9, of which head connects to the optical fiber sensing experimental instrument placed in the micro bend deformer. Using vernier adjust knob (spiral micrometer, minimum scale of 10^{-5} m) in the first place to contact with the fiber of microbend device. Record the PIN detection signal after obtaining amplification output voltage value. At the same time record the current value of the spiral micrometer, then record output value of voltage output in every 20 μm. Note: do not put force on optical fiber in case it is broken. Draw a curve of data, and the curve can be seen as a micro displacement measurement calibration curve, used for micro displacement detection. Using this curve can be convenient to study the characteristic of optical fiber optic microbend loss.

Analysis and ponder

How to make use of optical fiber sensing experimental device with micro bending plate, according to the need to design the experiment to realize the pressure test. How to obtain scale factor before conducting pressure measurement?

Optical Communication Experiment

1. Background and application

In ancient times, people exchanged information through brief language, murals, etc. Since hundreds of years, people have transferred information through languages, icons, fireworks, bamboo slips, paper and books, etc., examples of which are flames and smokes, transferring information by pigeon, mail delivery by post-horse, etc. (see Fig. 17-1). Even now, some primitive tribes in some countries still preserve the ancient communication modes such as stroking drums and sounding the bugle.

Optical communication is the communication with light wave as the carrier wave. The media of optical communication used are air, water, liquid fiber tube, glass fiber, optical fiber cable, and even outer space is under attempts; the wavelength used for optical communication ranges from infrared light, visible light to high frequency ray. However, nowadays optical fiber is dominant in optical communication field. Reviewing the development history of optical communication, there are approximately three stages for people utilizing light to transfer information:

(1) Visual communication

The following items all belong to communication signals of visible light: the alarming of the beacon tower in ancient China, transferring information by flag signal in Europe, the signal trees and signal pistol cartridge during the war, the beacon tower used by ships, signal lamps, etc. (see Fig. 17-2).

(2) Laser atmospheric communication

In 1880, American Bell invented the "phototelephone" transferring sounds with optical wave as the carrier wave. Bell's phototelephone is the prototype of modern optical communication. In 1960, American Maiman invented the first ruby laser, bringing new dawn to optical communication. The invention and application of lasers made optical communication which had slept deeply for 80 years step into a new stage. People did a plenty of experiments of laser atmospheric communications, however, the study of laser atmospheric communication which was ever prosperous in mid 1960s stayed at the same level forever, since there is severe absorption and scattering in the atmosphere and it is greatly influenced by the changes of weather.

Fig. 17 – 1   Transferring information by flame

Fig. 17 – 2   Communication by flag signal

(3) Optical fiber communication

In 1966, Chinese-born American Dr. Charles Kuen Kao (see Fig. 17 – 3) firstly proposed the concept of ultra low loss fiber with the principles of radio waveguide communication. Dr. Kao foresaw that as long as getting rid of various impurities in glass, the absorption of light can be greatly reduced and practical ultra low loss fiber can be produced. After many researches and experiments, American Corning Incorporated firstly managed to research and develop the quartz optical fiber with loss of 20 dB/km. The diameter of this optical fiber is as thin as human hair, and soft to circle. In the same year,

Fig. 17 – 3   Charles Kuen Kao

GaAlAs heterostructure semiconductor laser realized the continual working at room temperature and provided the ideal light source for optical fiber communication. Since then, optical fiber communication has greatly developed. In 1976, American Bell Laboratory established the first practical optical fiber communication line in the world from Atlanta to Washington D. C., in which the speed is 45 Mb/s, the fiber is multimode, the light source is LED and the infrared light wavelength is 0.85 um. Since then, the transmission loss of optical fiber has been decreasing, the bandwidth has been increasing, the performance of light source and optical detectors has been developed and the life of them has been increasing. In recent 20 years, different kinds of optical fiber communication systems and optical fiber networks has established like the bamboo shoots after a spring rain, which indicates the mighty competitiveness of optical fiber communication.

There are several advantages of optical fiber communication:

(1) The volume of the communication is large; the distance of transmission is long;

(2) It can not be disturbed by electromagnetic field, and thus can work in strong electric field;

(3) It has a strong corrosion resistence, and can work in the environment of harmful gas, such as chemical mines, etc.;

(4) The weight of optical fiber is light, and can be installed on military equipments such as planes, etc., thus it can reduce load, increase its speed and performance;

(5) The sources of the raw and processed materials are abundant, and it can save lots of metals.

Optical fiber communication has been widely used in modern world. It is not only widely used on earth, but also much more used for over-ocean communications(see Fig. 17-4). The submarine optical fiber cable line has been constructed across the Atlantic Ocean and the Pacific Ocean. Optical fiber nearly surrounds the whole earth. Optical fiber communication is also widely used in local network, LAN, MAN and ISDN, transferring non-voice service such as HDTV, data, etc., and creates many opportunities for people to fully enjoy a rich and colorful information service.

(a)   (b)   (c)

Fig. 17-4   kinds of the optical fiber cable

(a) The construction of aerial optical cable; (b) The laying of underground optic fiber cable; (c) The laying of submarine optical fiber cable

2. Experiment principles

The components of optical communication system

The components of optical communication system are shown in Fig. 17 – 5.

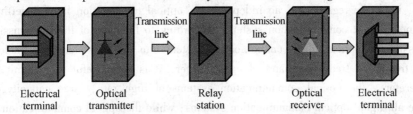

Fig. 17 – 5   The components of optical communication system

(1) Optical transmitter

Optical transmitter is the optical transceiver of photovoltaic conversion (electrical to optical). It consists of two parts-driving circuit and light source. It can convert the electrical signal from electrical terminal into the optical signal and then couple the optical signal into the transmission line.

(2) Optical receiver

Optical receiver is the optical transceiver of photovoltaic conversion (optical to electrical). It also consists of two parts: optical detection circuit and amplifier circuit. It can convert the optical signal from the transmission line into electrical signal through the optical detection device and then amplify the faint electrical signal to an adequate level by amplifier circuit and send it to the receiver.

(3) Transmission line

In a communication system, the transmission line often consists of optical fiber and optical cable. However, in the outer space, the laser transmission is not impacted by the atmosphere, so there are still enormous prospects to use laser as space communication.

(4) Relay station

Relay station is also called repeater station. The optical transmission system with replay station is called communication relay. The replay station has two functions: compensating the optical attenuation and reshaping the distortion pulse of the waveform. The traditional optical replay station adopts the O – E – O model. The photoelectrical detection device converts the faint and anamorphic optical signal transmitted by optical fiber into electrical signal, and then, by amplifying, reshaping and timing, return the electrical signal to the original electric pulse. After that, convert the electric signal into optical signal and send the optic pulse to next transmission line. Since the EDFA was invented, optical relay has realized all-optical relay. And this technique is still a hot topic in the communication realm.

(5) Optical fiber connector, coupler, and other passive devices

The connection of optical fiber, the connection and coupling of optical fiber and optical transceiver need optical fiber connectors and couplers. When the above system is added with the proper port, it can be regarded as an independent "optical transmission line" inserting into the existing or newly-built communication system. According to the type of the transmission signal, we can classify the optical fiber communication system into digital system and analog system. Because of the broad frequency band of optical fiber, it is very useful to transmit the digital signal. Therefore, the optical communication systems of high rate, large capacity, and long distance are all digital optical communication systems; while the optical communication systems of short distance and small capacity are all analog optical communication systems.

The characteristics of light source

There are mainly two kinds of light source in communication: LED and LD(laser diode).

LED is spontaneous radiation luminescence depending on the recombination of electron-hole pairs near P - N junction. When electrifying a positive voltage on the P - N junction of LED, electric field will weaken the built-in electric field, narrow the space charge region and strengthen the carriers' diffuse movement. Since the electron mobility is always much greater than hole mobility, the diffusion of electrons from N section to P section is the main part of diffuse movement of carriers. Considering the energy band theory of semiconductor, when electrons in conduction band recombines with the hole in the valence band, electrons transit from top level to low level and emit the surplus energy by the form of emitting photon to produce the electroluminescent phenomenon. This is the luminescence theory of LED.

LD-luminesces by stimulated emission. It has a high power, narrowing spectral width, stable wavelength and long life output. However, it is expensive. It applies to the transmission system of large capacity and long distance.

We adopt the LED source in this experiment, which has no threshold current. When there is a large current, the relationship between the drive current I and the output power P is non-linear. When transmitting the analog signal, we must know the linear output region of LED source and take feeded-back action in non-linear region.

The transmission of direct intensity modulation (DIM) of analog signal

Direct intensity modulation (DIM) refers to directly modulate the intensity of light source through the baseband signal, i.e. make the light intensity of light source change with the transferring signal. At this time, the transmission bandwidth of the optical fiber communication satisfies the bandwidth of the signal, the disadvantage of which is the high demand of a high level linearity of light source. If we use the ordinary LD as the light source, it can't achieve a proper performance because of the limitation of the nonlinearity of the light source, pattern noise and mode partition noise. In this modulation method, we often use the LED as the light source which has a better linearity and uses drive current-output characteristics. Fig. 17 - 6 shows the framework of direct intensity modulation in optical fiber transmission systems.

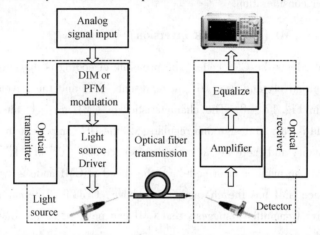

Fig. 17 - 6 The framework of DIM optical fiber transmission

The transmission of pulse frequency modulation (PFM) of analog signal

PFM is a high quality analog signal transmission way in recent years, which is a pretreatment process before modulating the signal. After PFM, we can prevent the influence caused by the nonlinearity of the light source and exchange it for the improvement of transmission quality.

There are two ways of PFM: one is modulating the repetition frequency of pulse changing with the signal amplitude linearly with the same pulse width; the other is that pulse duty cycle is 1 : 1, but the modulation of repetition frequency of pulse changes proportionally with signal amplitude, which is also called square wave modulation. In this part, we mainly discuss the latter one, and Fig. 17 - 7 shows the scheme of PFM.

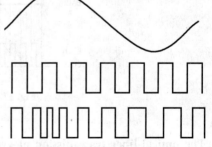

Fig. 17 - 7 The scheme of PFM

The transmission of digital coding

In the optical fiber digital transmission, the digital signal from the electrical terminal can not be transmitted directly in general, but by the code pattern transformation, it can be transformed into the optical code pattern that is appropriate to the optical fiber digital transmission system. Thus the choice of the transmission code pattern is an important issue under consideration.

The decoding rule of CMI (Code Mark Inversion)

This experiment adopts the CMI code. The principle of coding is that: use "00" and "11" to represent "1" alternatively and use "01" to represent "0", and the waveform of coding time sequence is shown in Fig. 17 – 8. The characteristics of CMI are: ① the encoding-decoding circuit is simple, and is easy to design and regulate; ② the DC component in the power spectrum of CMI code is constant, and there are at most three successive "0" or "1" in a code, so it is easy to sample; ③ it can monitor error code. ④ the speed of CMI code is twice faster than that before encoding. Since CMI has the characteristics above, CCITT (International Telephone and Telegraph Consultative Committee) suggests that CMI can be used as the line transmission code when the speed of the optical fiber transmission system is below 8,448 kb/s. CMI code is the inversion code of signal, a kind of 2-electric-level nonreturn-to-zero code.

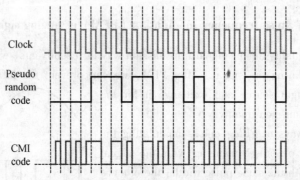

Fig. 17 – 8 The waveform of coding

The optical fiber transmission of visual signal

Fig. 17 – 9 gives us a flow chart of optical fiber analog video signal transmission system. It consists of a transmitter, a long optical fiber and a receiver. In the transmitter, the changes of input signal voltage is converted into the proportioned changes of LED's current near the bias point. Since the light intensity of LED's output changes linearly with the current, the intensity changes produced by this are the analog mode of the input signal. Light enters one end of the optical fiber and propagates into the receiver. At the receiver, the light from the end of optical fiber irradiates LED. The receiver turns the changes of light intensity into the proportioned current changes. Finally, the amplifier turns the weak current into the voltage changes which has the same amplitude as the original input signal. Many ways of modulation can be used to send video signals to optical fiber lines. The easiest modulating method is analog base band modulation. By this method, the video signal is returned to the transmitter and restored in the receiver.

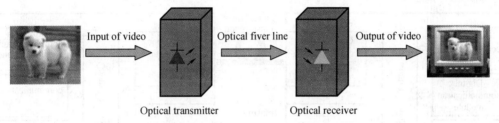

Fig. 17-9 The flow chart of video signal transmission system

3. Experiment purposes

Through this experiment, understand some basic knowledge about optical communication system; learn the structure and characteristics of LED light source; master the basic principles of DIM and PFM of analog signal; understand the basic components of digital optical fiber communication, and observe the phenomenon of optical fiber video transmission.

4. Experiment instruments

Optical communication experiment system:
(1) Optical communication experiment system
The experiment system and its scheme are shown as in Fig. 17-10 and Fig. 17-11:
(2) Optical transmitter of the experiment
The panel of the optical transmitter is shown as in Fig. 17-12, among which the functions of control region are:
Light source access: The switch of light source; after pressed down, the light source will be

linked to drive circuit.

Fig. 17-10   Optical communication experiment system

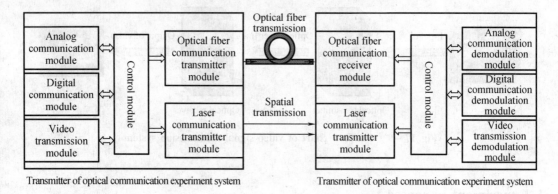

Fig. 17-11   The principle diagram of optical communication experiment system

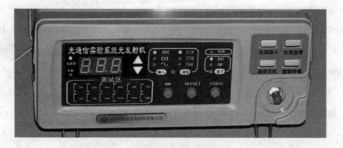

Fig. 17-12   The panel of transmitter

Light source option: The toggle switch of LD and LED.

Communication mode: The toggle switch of analog communication and digital communication. When pressed down, it becomes digital mode.

Video transmission: The video switch; when pressed down, it becomes video transmission mode; meanwhile, shut down other communication modes.

Optical Communication Experiment — Experiment 17

LED: The numbers displayed on it shows the bias current of luminescent device, and the unit is mA.

The upper triangular button and lower triangular button: The regulating button of bias current. The upper triangular button increases the bias current and the lower button decreases the current. A single trigger makes 1mA step each time, and when keeping it being pressed down, there will be a continuous regulating.

Input: The selection button of the input mode. There are three modes: MIC (Voice signal input), EXT (Extra signal input) and Sine Wave. The green light indicates the current mode.

Modulation: The selection button of the modulating mode. There are three modes: DIM, PFM and PWM. The green light indicates the current mode.

Digital: The selection button of digital communication mode. There are two modes: DAT and CMI. When it works with DAT mode, it transmits the pseudo-random code data, and the code pattern is 000011101100101; when it works with CMI mode, it transmits the data after coding the pseudo-random code by CMI.

The functions of each test point in test region:

TP1: The input waveform of video input device; TP2: Extra waveform; TP3: The sine waveform produced by the device itself; TP4: The output waveform of PFM modulating unit; TP5: The output waveform of PWM modulating unit; TP6: The clock of digital transmission system (512K); TP7: The pseudo-random code produced by the device itself; TP8: The waveform after coding of CMI; TP9: Video signal waveform; TP10: Signal waveform of light source.

(3) The receiver of optical communication experiment system

The panel of optical receiver is shown as in Fig. 17 – 13, among which the functions of control region are:

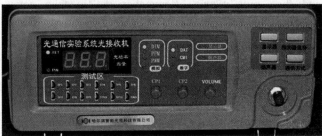

Fig. 17 – 13   The panel of receiver

The detector selection: The toggle switch of PIN and PET.

Communication mode: The toggle switch of analog and digital communication; when pressed down it becomes digital mode.

Display: The switch of video function; when pressed down, the video signal will be switched on the screen.

Loudspeaker: The switch of loudspeaker.

LED: The numbers displayed on it shows the relative value of optical power detected by optical detector.

Analog: The selection button of the analog mode. There are three modes: DIM, PFM and PWM. The green light indicates the current mode.

Digital: The selection button of digital communication mode. There are two modes: DAT and CMI. When it works with DAT mode, it transmits the pseudo-random code data, and code pattern is 000011101100101. When it works with CMI mode, it transmits the data after the decoding of CMI.

The functions of each test point in test region:

TP1: The detector's output after preamplification; TP2: The output waveform of DIM demodulation unit; TP3: The output waveform of PFM demodulation unit; TP4: The output waveform of PWM demodulation unit; TP5: The video signal after demodulating; TP6: The digital signal after truing; TP7: Inverse pulse sequence departed by clock recovery circuit; TP8: The clock signal departed by clock recovery circuit; TP9: The clock after truing; TP10: Data after decoding of CMI.

The connection of light source, detector and optical fiber

Aim the location bulge of the optical fiber connector at the light source or the notch of the detector; move onwards and insert the ceramic core into the base; then rotate it clockwise. When cutting off the connection, push the connector onwards slightly, rotate the connector counterclockwise, and then pull it out. Remember not to take action at random (see Fig. 17 – 14).

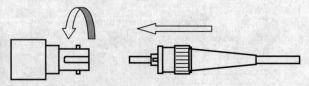

Fig. 17 – 14　The optical fiber connector

Digital storage oscilloscope

Digital storage oscilloscope is an advanced oscilloscope which firstly samples the input signal and makes A/D conversion, and then turn the input analog signals into digital signals and store them in the memory. The microprocessor of the oscilloscope turns the stored digital signal into visual waveform. Because of its memory function, this kind of oscilloscope is particularly appropriate to capture and display the single pulse signal. Of course, it can display periodic singal stably. However, it can not scan continuously like analogous oscilloscope. After sampling, it will do some treatments, such as A/D conversion, storing, displaying, etc. After the work, it will sample the signal again. The regular digital storage oscilloscope only uses 1% time to capture signal. Since the oscilloscope turns the waveform into a digital pattern, it is easy to do various calculations, such as using the FFT (fast fourier transform) to do spectrum analysis, automatically measure the peak-peak value, rising time, etc. It can also be linked to other computers. Because of these excellent advantages, the digital storage oscilloscope is widely used in measurement realm.

Now, let's briefly introduce the digital storage oscilloscope (TDS1002) of Tektronix company, the appearance of which is shown in Fig. 17 – 15. The front panel can be divided into five parts according to different functions: display region, vertical control region, parallel control region, trigger region and function region. And there are five menu buttons and three input ports. Each button's name on the panel, the operation method and basic function are shown in Table 17 – 1.

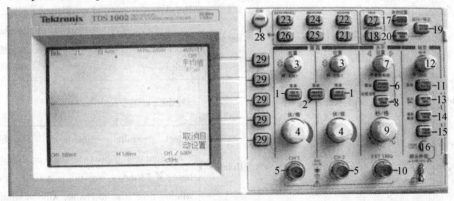

Fig. 17 – 15 The appearance of digital storage oscilloscope of TDS1002

1—CH1, CH2 menu; 2—Math calculation menu; 3—the positions of cursor 1 and cursor 2; 4—voltage/grid (CH1, CH2); 5—input port; 6—parallel menu; 7—parallel position; 8—set-to-zero; 9—second/grid; 10—external trigger; 11—trigger menu; 12—triggering level; 13—set-to-50%; 14—enforced trigger; 15—trigger view; 16—the probe compensation; 17—the automatic settings; 18—the default settings; 19—operate/stop; 20—single sequence; 21—display menu; 22—collecting menu; 23—menu of save/move stock; 24—measuring menu (auto); 25—cursor menu (manual); 26—auxiliary function menu; 27—helping menu; 28—Printing; 29—the selection button of five menus

Table 17-1(a)   The button's names, operation method and basic functions of TDS1002

System	Numbering	Name	Operation method	Main functions
Vertical control	1	CH1, CH2 menu	Pressed down	Display the options of vertical menu and open/close the display of channel's waveform
	2	Math calculation menu	Pressed down	Display the calculation of the waveform and can be used to open/close the mathematical waveform
	3	Positions of cursor 1 and cursor	Rotate	Orientate the waveform vertically
	4	Voltage/grid (CH1, CH2)	Rotate	Select the calibration coefficient
	5	Input (CH1, CH2)	Link	Input the signal
Parallel control	6	Parallel menu	Pressed down	Display the "parallel menu"; use the selection button to set the time base mode
	7	Parallel position	Rotate	Regulate the parallel positions of all channels and mathematical waveforms; the resolution of this control change with the different settings of time base
	8	Set-to-zero	Pressed down	Set the parallel position to zero
	9	Second/grid	Rotate	Select parallel time/grid (calibration coefficient) of main time base or window time base; if "window region" is activated, change window time base to change window width
	10	External trigger	Link	Input external trigger signal
Trigger control	11	Trigger menu	Pressed down	Display the "trigger menu"; use the selection button to set the trigger mode

Optical Communication Experiment **Experiment 17**

Table 17-1(b)

System	Numbering	Name	Operation method	Main functions
Trigger control	12	Triggering level	Rotate	When we use "edge" to trigger, the basic function of the "level" button is to set level amplitude, and in order to collect the signal, the signal amplitude must be higher than level amplitude. We can also use it to implement other functions of "user option". The LED under the button will brighten up to indicate the corresponding functions
	13	Set-to-50%	Pressed down	Set the triggering level as the vertical middle point of peak value of trigger signal
	14	Enforced trigger	Pressed down	No matter whether the trigger signal is appropriate, the collection of signal will be completed. If the collection has stopped, the button does not impact the result
	15	Trigger view	Pressed down	When this button is pressed down, what it displays is not channel waveform but trigger waveform. This button can be used to check the impacts of trigger settings such as trigger coupling on trigger signal
	16	The probe compensation	Pressed down	Check whether the connection compensation is correct rapidly
Observation control	17	The automatic settings	Pressed down	Automatically set the control state of oscilloscope to produce graphic display applicable to output signal
	18	The default settings	Pressed down	Bring up the factory Settings
	19	Operate/stop	Pressed down / Pressed down again	Continuously collect waveform or stop he collection

253

Table 17 – 1 (c)

System	Numbering	Name	Operation method	Main Functions
Observation control	20	Single sequence	Pressed down	Collect a single waveform, and then stop
	21	Display menu	Pressed down	Display the display menu
	22	Collecting menu	Pressed down	Display the collecting menu
	23	Menu of save/move stock	Pressed down	Display the menu of save/move stock of settings and waveforms
	24	Measuring menu (auto)	Pressed down	Display the measuring menu
	25	Cursor menu (manual)	Pressed down	Display the "cursor menu". When it displays the "cursor menu" and the cursor is activated, the control method of "vertical position" can regulate the cursor's position. When this menu is shut down, the cursor keeps being displayed (unless the "type" option is set to "closed"). But it can not be regulated
Auxiliary application	26	Auxiliary function menu	Pressed down	Display the "Auxiliary function menu"
	27	Helping menu	Pressed down	Display the "helping menu"
	28	Printing	Pressed down	Printing operations
	29	The selection button of five menus	Pressed down	Cooperate with the menus, and select the parameters or functions

5. Experiment content and operation key points

I – P Characteristic research of LED light source

This instrument adopts LED light source, the center wavelength of which is 850 nm. In order to obtain the I – P characteristic curve of LED, do the experiment according to the following steps:

(1) Take an optical fiber matching the device. Link one end of the fiber with the socket of LED light source, and the other with the socket of PIN detector.

(2) Press the "communication mode" to set the mode of optical transmitter and receiver as analog communication mode (set the "video transmission" key open).

(3) Use the "modulation" button to set the modulating mode as "DIM" of the transmitter. Use the "analog" button to set the demodulating mode as "DIM" of the receiver. Now, the optical communication experiment system is working with DIM mode.

(4) Regulate the "input" on optical transmitter to "MIC", press down the "light source access" button on optical transmitter, and use the "light source option" button to select LED light source. Use the "detector option" button of the transmitter to select the PET detector.

(5) Regulate the bias current button (▲ and ▼) on optical transmitter, and increase the drive current from 0 mA (each time increase 1 mA). Write down the corresponding mechanical equivalent of light in Table 17 – 2 until there is no obviously changes (No more than 70 mA).

(6) Decrease the bias current of light source near zero and prepare for the next experiment.

Notes: Bias current and mechanical equivalent of light are not real value but relative value, and are linear with the real value. And there will be slight differences with different devices.

Direct intensity modulation transmission of analog signal

(1) Take an optical fiber matching the device. Link one end of the fiber with the socket of LED light source, and the other with the socket of PIN detector.

(2) Press the "communication mode" to set the mode of optical transmitter and receiver as analog communication mode.

(3) Use the "modulation" button to set the modulating mode as "DIM" of the transmitter. Use the "analog" button to set the demodulating mode as "DIM" of the receiver. Now, the optical communication experiment system is working with DIM mode.

(4) Regulate "input" of the optical transmitter to "sine wave", press down "light source

access" button of the transmitter, and use the "light source option" button of the transmitter to select LED light source and the "detector option" button of the receiver to select the PET detector.

(5) Measure the signal original waveform through the test point TP2 on the panel of transmitter, and this signal is signal 1. Write down the frequency and peak-peak value of sine wave in the Table 17 – 2.

(6) After optical fiber transmission, observe the waveforms which are directly detected by detector and amplified by amplifier. Now the waveforms are not integrated sine wave.

(7) Increase the bias current of light source of the transmitter to make the signal received from the TP2 just distort (when the waveform turns into integrated sine wave). Write down the bias current, i.e. the bias current corresponding to no distortion.

(8) Use the AM button on transmitter to increase the amplitude of signal 1 (with a greater increase) until the screen shows the signal with distortion again demodulated by receiver. This signal after changing is written as signal 2. Write down the frequency and amplitude of signal 2 and find the corresponding bias current of no distortion.

The PFM transmission of analog signal

(1) Take an optical fiber matching the device. Link one end of the fiber with the socket of LED light source, and the other with the socket of PIN detector.

(2) Press the "communication mode" to set the mode of the optical transmitter and receiver as analog communication mode.

(3) Use the "modulation" button to set the modulating mode as DIM. Use the "analog" button of the receiver to set the demodulating mode as PFM. Now, the optical communication experiment system is working with PFM mode.

(4) Regulate the "input" on optical transmitter to "MIC", press down the "light source access" button on optical transmitter, and use the "light source option" button to select LED light source. Use the "detector option" button of the receiver to select the PET detector.

(5) Observe the TP4 on transmitter and write down the frequency of waveform which is now the centre frequency of PFM.

(6) Set the input signal of transmitter as "sine wave", and observe the waveforms of TP1 and TP3, which are the waveform of direct output of the detector and the waveform demodulated by detector, respectively.

(7) Change the bias current of the transmitter and observe the waveforms of TP1 and TP3. Think about whether the current changes have an obvious impact on transmission quality.

Optical Communication Experiment — Experiment 17

Digital coding transmission experiment

(1) Take an optical fiber matching the device. Link one end of fiber with the socket of LED light source, and the other with the socket of PIN detector.

(2) Press the "communication mode" to set the optical transmitter and receiver as digital communication mode. Use the "digital" button to select "CMI", and now the outputs of the transmitter are the data coding by CMI.

(3) Use oscilloscope to observe the output system clock signal of TP6.

(4) Use oscilloscope to observe the TP7, and write down the waveform of this point which is 15 digits pseudo-random code produced from the pseudo-random code generator in the transmitter.

(5) Use oscilloscope to observe the CMI code from TP8.

(6) Link the oscilloscope with TP1, and observe the digital signals transmitted by optical fiber and amplified by photoelectric detection.

(7) Use oscilloscope to observe the digital signal of TP6 after shaping.

(8) Cut off the optical path and observe the TP7, TP8 and TP9. At this time, the signals are the clock signals picked up by the clock with no outer trigger signal.

(9) Recover the optical path, and TP10 is the circuit after decoding. Observe and write down the code pattern at this point.

The optical fiber transmission experiment of video signal

(1) Take an optical fiber matching the device. Link one end of the fiber with the socket of LED light source, and the other with the socket of PIN detector.

(2) Press down the "light source access" button on optical transmitter, and use the "light source option" button to select LED light source. Use the "detector option" button of the receiver to select the PET detector.

(3) Use the "video transmission" button on transmitter and "screen" button on receiver to set the communication method as video transmission mode, and link the camera with the screen.

(4) Regulate the bias current button (▲ and ▼) on optical transmitter, and observe how the different bias currents influence the video signal transmission.

(5) Regulate the bias current and "VIDEO" button to obtain the clearest video signal on the screen.

After the experiment, make sure to release the "light source access" button on the transmitter. Shut down the device and tidy the experiment devices.

NOTE:

(1) Pay attention to protect optical fiber in the experiments, avoiding pulling or pressing the optical fiber forcefully. The bending degree of the fiber can not be too big to avoid breaking the fiber.

(2) When not using the fiber, cover good dust cap on it.

(3) When not using the device, cover good dust cap on the light source and the detector.

(4) There is an over current protection circuit in optical source drive part, but in case of accident, do not get an extremely high bias current in the experiment.

(5) Choose the appropriate bias current. Working long hours in the great current will affect the service life of the light source.

6. Data recording and processing

I – P characteristic research of LED light source

Table 17 – 2 The relationship between drive current I and the output of optical power P

Current/mA	0	1	2	3	4	5	6	7	8	9
Optical power equivalence										
Current/mA	10	11	12	13	14	15	16	17	18	19
Optical power equivalence										
Current/mA	20	21	22	23	24	25	26	27	28	29
Optical power equivalence										
Current/mA	30	31	32	33	34	35	36	37	38	39
Optical power equivalence										
Current/mA	40	41	42	43	44	45	46	47	48	49
Optical power equivalence										
Current/mA	50	51	52	53	54	55	56	57	58	59
Optical power equivalence										
Current/mA	60	61	62	63	64	65	66	67	68	69
Optical power equivalence										

(1) Draw up the I – P characteristic curve of LED light source on a coordinate paper.

(2) Analyze the I – P characteristic curve of LED light source and point out the area with better linearity.

(3) If using this device to transmit signal, how to select the proper static working point, And in which range should the signal amplitude be kept?

Direct intensity modulation transmission of analog signal

Table 17 – 3   The experiment data of DIM transmission of analog signal

Signal	Frequency/Hz	peak-peak value/V	Non-distortion bias current/mA
Signal 1			
Signal 2			

(1) Point put the bias current I_1 and I_2 on the I – P characteristic curve.

(2) Briefly introduce the characteristics of DIM.

The PFM transmission of analog signal

(1) Present the center frequency of PFM produced by the device.

(2) Present the frequency of the sine wave produced by the demodulation of the receiver, and compare it with the original waveform and figure out the experiment conclusion.

(3) Briefly introduce the characteristics of PFM.

Digital coding transmission experiment

(1) Present the pseudo-random code in the experiment.

(2) Present the data after coding by CMI in the experiment

7. Analysis and questions

(1) This instrument adopts LED light source, the center wavelength of which is 850 nm. Can we see the light produced by LED?

(2) Why do we obtain the I – P characteristic curve of LED light source, How to select the proper static working point, and in which range should the signal amplitude be kept?

(3) In DIM of analog signal, if the static point is not at the linear interval of the light source, what impact will occur for the transmission consequence? In terms of PFM of analog

signal, if the static point is not at the linear interval of the light source, what impact will occur for the transmission consequence?

(4) Why do we use the coding transmission?

(5) Which parameters will influence the transmission quality?

Experiment 18

Modification and Calibration of Electricity Meters

1. Background and application

Wattmeter is one of the most basic electricity measurement tools, which can be divided into DC wattmeter, AC wattmeter, AC – DC wattmeter according to operating current; ammeter and voltmeter according to application; and pointer wattmeter and digital wattmeter according to the ways of reading. Commonly used wattmeters include DC ammeter, AC ammeter, DC voltmeter, AC voltmeter, ohmmeter and avometer, etc. All these wattmeters can modified from galvanometer (popular name is meter).

The development of Wattmeter, as a measuring instrument, is closely related to the development of electromagnetics theory and continuous improvement of experiment level. Since the first electroscope of the world (in 1743), movable coil galvanometer (in 1836), Wheatstone bridge (in 1841) and DC potential difference meter (in 1861), etc. came out, electromagnetic measuring instrument had gradually developed from development phase in laboratory to commercial products by 1930s. Classical electrician instrument has been finalized basically in design theory and process structure. In 1960s, the precision of wattmeter has greatly improved; the accuracy rate of a series of wattmeter has achieved the level of 0.1. Since then, the classic instrument has basically stayed at the level due to the limits of material and technology.

The ammeters or voltmeters used in laboratory are generally magnetic-electric wattmeters, which have the merits such as high sensitivity, low power consumption, strong anti-effect of ambient magnetic field, universal scale, and convenient reading, etc. In general, the meter of unmodified wattmeter only permits the passing of microampere current because of high sensitivity and very low full scale current (voltage); therefore, it can only measure very low current or voltage. If it is used to measure higher current and voltage, modification must be carried out to amplify the range of measurement, and this process is called range multiplication of wattmeter. Any instrument (especially self-built instrument) should be calibrated before using, especially before precision measurement. Therefore, calibration is a very important technique in experiment technique. In accordance with national standard (GB/T 7676.2—1998), ammeter and voltmeter should be classified according to the following class index expressed accuracy class: 0.05, 0.1, 0.2, 0.3, 0.5, 1.0, 1.5, 2.0, 2.5, 3.0, 5.0, altogether 11 classes.

2. Experiment purposes

Know the working principle of wattmeter and the function and operation of common basic electricity instrument; master how to design simple wiring diagram to find out the solutions to the problem; learn how to determine accuracy class of wattmeter.

3. Design Requirements

It is required that design be reasonable, principle be correct, proper instrument be selected, circuit design be correct and the layout be reasonable.

4. Experiment instruments

The modified wattmeter is shown in Fig. 18 – 1, the power supply is shown in Fig. 18 – 2, slide rheostat is shown in Fig. 18 – 3, resistance box is shown in Fig. 18 – 4, standard milliamperemeter is shown in Fig. 18 – 5, and standard voltmeter is shown in Fig. 18 – 6.

Note the electric poles of the modified wattmeter: Red is positive and black is negative. When wiring, pay attention that current flows in from positive pole and flow out from negative pole.

For HY1711 – 3S Dual-channel output trackable DC regulated power supply, only one channel of it is used in the experiment, and either the left channel or the right channel can be used. Note that when using the left channel, left channel voltage regulating knob should be adjusted; when using the right channel, right channel voltage regulating knob should be adjusted.

Modification and Calibration of Electricity Meters — Experiment 18

Fig. 18 – 1 Modified wattmeter

Fig. 18 – 2 HY1711 – 3S Dual-channel output trackable DC regulated power supply

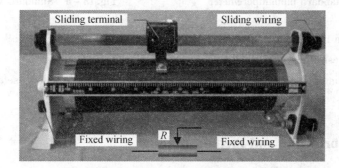

Fig. 18 – 3 Slide rheostat

The experiment uses the voltage dividing function of slide rheostat. Note that 3 terminals of slide rheostat should all be used: Two fixed terminals are connected to the positive and negative poles of the power supply, and the sliding terminal is connected to the circuit of modified wattmeter; and where the sliding terminal is placed before turning on the power.

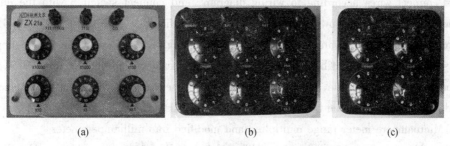

(a)　　　　　　　　　　(b)　　　　　　　　　　(c)

Fig. 18 – 4 A few common rheostats

(a) ZX21a Multirange resistance box; (b) ZX21 Multirange resistance box;
(c) Single range resistance box

Fig. 18 – 5 Standard milliamperemeter

Fig. 18 – 6 Standard voltmeter

The maximum range of rheostat used in the experiment is $0 \sim 99,999.9\ \Omega$. It should be noted that initial resistance vale should be preset when using.

Standard milliamperemeter is AC – DC wattmeter: For the experiment, select 10 mA range. Standard voltmeter: For the experiment, select 3 V range.

5. Experiment principle and method prompt

(1) The measurement of meter inherent resistance

The microampere meter (galvanometer) used for modification is usually called meter, and coil resistance R_g inside the meter is called meter inherent resistance. If the microampere meter is modified for multiplication range, first measure meter inherent resistance. The experiment uses substitution method to measure wattmeter inherent resistance, whose circuit is shown in Fig. 18 – 7.

First connect wires according to wiring diagram, and connect switch S with the end a to protect wattmeter. Turn on power, adjust slide end D of the slide rheostat R_1 to make G at full range (or a certain suitable value), and record the reading of G_0. Cut off the power supply E, connect switch S with the end b, first adjust R value of resistance box to about $3,999.9\Omega$, turn on the power, adjust R value again to make G_0 keep the original value, at the moment the indicated reading on resistance box R is meter inherent resistance R_g.

(2) Microampere meter range multiplied and modified into milliamperemeter

Full scale current of meter is very small, and is usually microampere quantity level. If the current over its range is required to be measured, its range must be multiplied. The method is that a shunt resistor R_p is parallel connected to both ends of meter as shown in Fig. 18 – 8. So most of current to be measured passes through shunt resistor and meter still keeps the maximum current I_g that is originally permitted to pass. In Fig. 18 – 8, the meter and shunt resistor R_p in dotted line is composed of a new range ammeter. Suppose the new meter range is I, which is known in accordance with Ohm's law

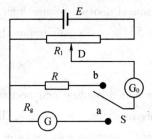
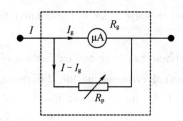

Fig. 18 – 7 Circuit for measuring microampere meter inherent resistance

Fig. 18 – 8 Schematic diagram of microampere meter range multiplied and modified into ammeter

$$(I - I_g)R_p = I_g R_g \qquad (18-1)$$

$$R_p = \frac{I_g R_g}{I - I_g} \qquad (18-2)$$

Suppose $\frac{I}{I_g} = n$, n is called multiplied multiple of wattmeter range, and shunt resistance is

$$R_p = \frac{R_g}{n-1} \qquad (18-3)$$

After meter parameters I_g and R_g are determined, in accordance with the multiple n required to multiply current range, the value of shunt resistance R_p parallelled can be found out to realize meter multiplied range and modification.

(3) Microampere meter range multiplied and modified into voltmeter

Full scale voltage of meter is very low, generally only a few tenths of volts. In order to measure higher voltage, a division voltage high resistor R_s is connected with the meter in series as shown in Fig. 18 – 9 to make the partial voltage, which is over the meter to bear, to drop to the resistor R_s. The integration composed of meter and division voltage high resistor R_s in dotted line is a voltmeter whose new range is U. It is known

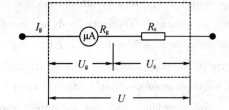

Fig. 18 – 9 Schematic diagram of microampere meter range multiplied and modified into voltmeter

according to Ohm's law that

$$U_s = I_g R_s = U - U_g \quad (18-4)$$

$$R_s = \frac{U - U_g}{I_g} = \frac{U}{I_g} - R_g \quad (18-5)$$

After meter parameters I_g and R_g are determined, the value of division voltage high resistance connected in series can be calculated from formula (18-5) according to required voltmeter range.

(4) Nominal error and calibration of wattmeter

Nominal error means the difference between wattmeter actual reading and accurate value, which includes the errors resulted in various imperfect factors in wattmeter structure. In order to determine nominal error, use the wattmeter and a "standard" wattmeter to measure the same current (or voltage) at the same time, which is called calibration. The result of calibration is to obtain the absolute error of each scale of wattmeter, and to select the maximum absolute error of them to be divided by new range of wattmeter, which is determined as the nominal error of the wattmeter, i. e.

$$\text{Nominal error} = \frac{\text{Max. absolute error}}{\text{Range}} \times 100\%$$

Note: "Maximum absolute error" in the formula is not the maximum difference of the value measured by modified wattmeter and the value measured by standard wattmeter. When calibrating, the selected standard wattmeter class is generally higher than the meter class of modified wattmeter, but the error of standard wattmeter sometimes should not be ignored. If the class of standard wattmeter selected is x_0 and the range is I_0, the maximum error of the wattmeter is $\Delta I_{0\max} = I_0 \times x_0\%$, and there must be $\Delta I_{\max} = |\Delta I_m| + \Delta I_{0\max}$, i. e. there is $\Delta I_{\max} = |\Delta I_m| + I_0 \times x_0\%$. Obviously, the class of selected standard wattmeter can not be less than that of modified wattmeter. If the class of standard wattmeter is the same as that of meter, and $\Delta I_{0\max}$ is not considered, wrong result that modified wattmeter class is higher than standard wattmeter class may be obtained. Therefore, only as specified in national test rules, when the error of standard wattmeter is less than $1/3 \sim 1/20$, compared with the meter error of calibrated wattmeter, the error of standard wattmeter can be ignored. If the condition can not be satisfied, the maximum absolute error of modified wattmeter must be calculated through $\Delta I_{\max} = |\Delta I_m| + I_0 \times x_0\%$.

In accordance with the measured result, draw out calibration curve (see Fig. 18-10), which can directly help us to analyze and judge whether the measured result is good or not. When wattmeter is actually used to measure current, the pointer of wattmeter may deflects from small to big and stops at a position, or maybe deflects from big to small and stops at a scale, and these two conditions appear randomly. Therefore, when calibrating a modified wattmeter, current must be calibrated from small to big, and repeatedly calibrated from big to small. In this way, for the same scale of the modified wattmeter, standard wattmeter may have two different indicated values, roughly processing can take its average value as accurate value of current in the circuit by using standard wattmeter, and then find out its difference value ΔI from the indicated number of modified wattmeter and draw the calibration curve. Find out the maximum value ΔI_{max} of them to determine the nominal error of modified wattmeter. Calibration curve takes ΔI (mA) as Y axis coordinate, and the indicated number I (mA) of multiplied and modified meter as X axis coordinate. Since each calibration position is obtained with standard wattmeter compared with the wattmeter to be calibrated, there is no function relation between them, so two adjacent calibration points are connected by a straight line, and the whole calibration line figure is fold line as shown in Fig. 18 - 10. When making figure specifically, coordinate division value should be marked (Note division value ratio and effective numbers).

According to the value of nominal error, wattmeter can be divided into different classes, called wattmeter accuracy rate. As specified in national standard (GB/T 7676.2—1998), ammeter and voltmeter should be classed according to the accuracy expressed to the following class index: 0.05, 0.1, 0.2, 0.3, 0.5, 1.0, 1.5, 2.0, 2.5, 3.0 and 5.0, all together 11 classes. For example, class 1.0, indicates the wattmeter nominal error is not higher than 1.0%, the others are by analogy. If wattmeter is calibrated, the nominal error obtained is not just the above values. According to the principle that error is taken in terms of the big, not the small, the class of wattmeter should be determined as low by one class: For example, after wattmeter is calibrated, the obtained nominal error is 1.8%, which is between 1.5 and 2.0, and the wattmeter should be determined as 2.0 class. Wattmeter class is usually marked on the meter of wattmeter. Note when using a wattmeter, the wattmeter of long-time service or after repairing should be used after calibration.

(5) Microampere meter modified into ohmmeter

Make a microampere meter, a high resistor R_0, a low resistor R_s, and power E compose a circuit as shown in Fig. 18 - 11. As resistor R_x to be tested has a corresponding relation with current I_g passing through microampere meter, the microampere meter is modified into an ohmmeter shown in the dotted line of the figure.

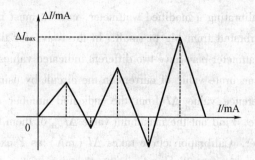
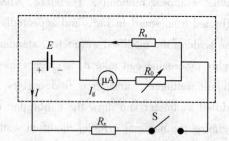

Fig. 18-10 Calibration curve

Fig. 18-11 Circuit of microampere meter modified into ohmmeter

Suppose the current passing through R_x is I, and the current passing through microampere meter is I_g; according to Ohm's law of total circuit:

$$I = \frac{E}{R_x + r + \frac{R_s(R_0 + R_g)}{R_s + R_0 + R_g}} \quad (18-6)$$

Where r is inherent resistance of battery, and R_g is inherent resistance of microampere meter. Since high resistance R_0 is mostly ten thousands ohms, then $R_s \ll (R_0 + R_g)$, $r \ll R_x$, and formula (18-6) can be approximate to

$$I \approx \frac{E}{R_x + R_s} \quad (18-7)$$

Suppose $R' = \frac{R_s(R_0 + R_g)}{R_s + R_0 + R_g}$, the obtained current passing through microampere meter is

$$I_g = \frac{IR'}{R_0 + R_g} \approx \frac{IR_s}{R_0 + R_g} \approx \frac{ER_s}{(R_0 + R_g)(R_x + R_s)} \quad (18-8)$$

As $I_g = k\alpha$ (α is deflection scale of microampere meter pointer), the above formula can become

$$\alpha = \frac{ER_s}{k(R_0 + R_g)(R_x + R_s)} \quad (18-9)$$

From formula (18-9), when E, R_s, R_0, R_g are fixed, R_x can be measured from pointer deflection scale of microampere meter

According to formulas (18-8) and (18-9), ohmmeter characteristics can be discussed:

①When $R_x = 0$, properly adjust R_0 to make microampere meter have full range, then

$$I_g = \frac{E}{R_0 + R_g} = I_{gm}$$

Where I_{gm} is the range of microampere meter.

When $R_x = \infty$, $I_g = 0$, microampere meter pointer points to zero scale. This shows that meter zero point (0 Ω scale) of ohmmeter is deflected to the most right, the end (i.e. the scale of $R_x = \infty$) of ohmmeter meter is deflected to the most left, which is opposed to the scale of meter of wattmeter.

②When $R_x = r + \dfrac{R_s(R_0 + R_g)}{R_s + R_0 + R_g} \approx R_s$, then

$$I_g = \frac{ER_s}{(R_0 + R_g)(R_x + R_s)} = \frac{1}{2}\frac{E}{R_0 + R_g} = \frac{1}{2}I_{gm} \qquad (18-10)$$

i.e. when R_x equals to the inherent resistance of ohmmeter, meter pointer is in the middle of scale, at the moment the resistance value is called mean value resistance of ohmmeter, i.e. $R_m \approx R_s$.

③According to formulas (18-8) and (18-9), it can be seen that R_x and $I_g(\alpha)$ are not a simple inverse relation, but:

When $R_x = 0$, $I_g = I_{gm}$

When $R_x = R_m$, $I_g = \dfrac{1}{2}I_{gm}$

When $R_x = 2R_m$, $I_g = \dfrac{1}{3}I_{gm}$

When $R_x = nR_m$, $I_g = \dfrac{1}{n+1}I_{gm}$

According to the above data, the drawn ohmmeter scale is shown in Fig. 18-12. It is known from the diagram that ohmmeter scale is not uniform: with the increase of resistance value, the scale is closer and closer.

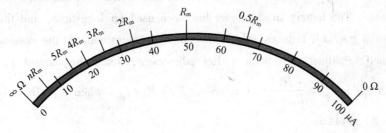

Fig. 18-12 Ohmmeter dial

④It is known from the above discussions that in principle ohmmeter only needs one range to measure any resistance value between 0 ~ ∞ Ω, but according to scale distribution condition, the more the resistance value to be measured is offset from R_m, the closer its scale is, and the more the reading error is. Therefore, when using a simple range ohmmeter to measure resistance value, only the resistance value near R_m is more accurate. If resistance values of different ranges are required to be measured, ohmmeter must have different ranges. The way is that mean value resistance $R_m \approx R_s$ is multiplied by $10^n (n = 0, 1, 2, \cdots)$.

As is shown in Fig. 18 – 13, for a range ohmmeter, whose mean value resistance is 10Ω, if the value is multiplied by 10 times or 100 times, each scale is increased by 10 times or 100 times, i.e. the range is increased by 10 times or 100 times. So it is known that R_s has the function of changing range. Note that the range of ohmmeter is only a relative concept, different from the concepts of ammeter and voltmeter ranges.

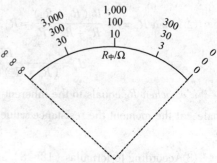

Fig. 18 – 13 R_m function

⑤It is known from equation (18 – 8) that for different ranges, when $R_x = 0$, the I_g values are not the same (Because R_s is different), so pointer position (zero position) varies a little with different ranges, which is not convenient for operation; therefore, the zero point corresponding to any range should be fixed. The solution to this problem is to adjust the value of R_0 to change the current ratio of two branch circuits to satisfy the requirements, so R_0 is called zero set resistor. The adjusting method is: Change the value of R_0 to make the pointer corresponding to $R_x = 0$ be offset to full scale, at the moment the current passing through microampere meter is just the maximum value I_{gm} it permits, and thus zero position is fixed.

Zero set resistor R_0 has another function to ensure meter scale not affected by battery electro-dynamic potential. The battery in ohmmeter has been used for long time, and thus its electro-dynamic potential gradually reduces, which also affects zero position. At the moment zero resistor R_0 can be used to adjust zero point. After adjustment, since the current passing through microampere meter is $I_{gm} = \dfrac{E}{R_0 + R_g}$, i.e. $E = (R_0 + R_m) I_{gm}$, when $R_x = 0$, substitute it into formula (18 – 8) to obtain

$$I_g = \frac{ER_x}{(R_0 + R_g)(R_x + R_s)} = \frac{I_{gm} R_x}{R_x + R_s} \qquad (18-11)$$

Therefore, adjust R_0 to make meter scale not affected by E (because in formula all is fixed except R_x).

Operation steps

(1) Connect circuit as shown in Fig. 18-11, where resistor box can act as R_s, R_0, R_x, R_s can use four-position simple range resistor box, R_0 and R_x use six-position multi-range resistor box, and power voltage is adjusted to about 1.5 V. Before turning on the power, first suppose $R_s = 100 \ \Omega$, $R_x = 0$, R_0 can be as high as possible (first can be adjusted to 19,999.9 Ω) to avoid burning microampere meter.

(2) After turning on the power, adjust R_0 to make microampere meter pointer deflect to full scale (corresponding to $R_x = 0$); in the experiment, R_s and R_0 are not changed any longer, because the range has been set, and zero point has already been adjusted properly.

(3) Change the value of R_x to make microampere meter pointer to the center scale, record the value of R_x at the moment (i.e. mean value resistance), adjust R_x value again in turn to the ohmage of 10, 20, 30, 60, 90, 120, 200, 280, 360, 480, 600, 1,000 and 2,000 till pointer deflects under one scale, record the deflection scale number corresponding to meter pointer, and fill them in Table 18-1.

Table 18-1 Data recording table for microampere meter modified into ohmmeter
$E = 1.5 \ V \quad R_s = 100 \ \Omega \quad R_{m=}(\ ) \Omega \quad R_0 = (\ ) \Omega$

R_x / Ω	10	20	30	60	90	120	200	280	360	480	600	1,000	2,000
Ohmage scale													

(4) Draw out the scale of the meter of ohmmeter.

(5) Change R_s into 10Ω, and modify it into another range ohmmeter.

6. Experiment content

(1) A 100 μA microampere meter is multiplied and modified into 10 mA milliamperemeter. Draw out designed circuit diagram, design a data table, draw out calibration curve, and determine the modified meter class.

(2) A 100 μA ammeter is multiplied and modified into a 3 V voltmeter. Draw out designed circuit diagram, design data table, draw out calibration curve, and determine the modified meter class (selected).

(3) Summarize the experiment, and write the experiment report.

7. Analysis and questions

(1) When calibrating ammeter, it is found that the reading of modified ammeter is a little high relative to the reading of standard ammeter; try to discuss how to adjust it to reach the value of standard ammeter, why?

(2) Is it possible to modify the wattmeter used for this experiment into a wattmeter with any range, for example: 50 μA or 0.1 V, why?

(3) If 0.5 A current is to be measured, which ampere meter below can be used to have the minimum error?

Range $I = 3$ A, and class $k = 1.0$; Range $I = 1.5$ A, and class $k = 1.5$; Range $I = 1$ A, and class $k = 2.5$.

What conclusion can be obtained from the comparisons among the results?

(4) After modified wattmeter are calibrated at zero point and full scale current, it is found that there are still errors at scales in the middle; analyze and discuss the reasons.

8. Appendix

The modification of multi-range wattmeter

(1) Multi-range DC ammeter

If a meter is modified into a multi-range ammeter, many shunt resistors must be connected in parallel to the meter. As shown in Fig. 18 – 14, changeover switch (called band switch) S is placed to different positions to obtain different ranges. The value of each shunt resistor can be found out by formula (18 – 3) in accordance to range requirements.

(2) Multi range DC voltmeter

If a meter is modified into a multi-range voltmeter, many divider resistors must be connected in series to meter. As shown in Fig. 18 – 15, changeover switch S is placed to different positions to obtain different ranges. The value of each divider resistor can be found out by formula (18 – 5) in accordance to range requirements.

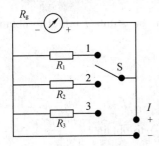
Fig. 18-14 Multi-range ammeter circuit

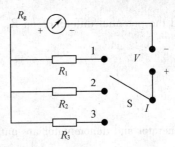
Fig. 18-15 Multi-range voltmeter circuit

The determination of characteristic curve

In the experiment, rheostat parameters (resistance and rated parameters) should be selected correctly. If the selection is suitable, the experiment is stable, precise and successfully; if the selection is not suitable, the experiment is not stable and even damages the equipment. The following is to analyze and research the voltage dividing characteristic curve and the shunt characteristic curve of slide rheostat.

(1) Voltage dividing characteristics curve of slide rheostat

The circuit of slide rheostat used as voltage dividing is shown in Fig. 18-16: Two fixed terminals (Terminal A and Terminal B) are separately connected to the positive and negative poles of the power, the voltage on load R_L varies with the motion of slide, and E is the end voltage of the power.

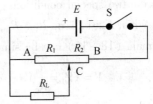
Fig. 18-16 Voltage dividing circuit

Now analyze the circuit: the total resistance R of the circuit can be considered as the integration that R_1 and R_L are connected in parallel and then connected in series with R_2, the voltage between Terminal A and Terminal B is V_0, and there is

$$R = R_2 + \frac{R_1 R_L}{R_1 + R_L} \quad (18-12)$$

Therefore, the total current I of the loop is

$$I = \frac{V_0}{R} = \frac{V_0}{R_2 + \dfrac{R_1 R_L}{R_1 + R_L}} \quad (18-13)$$

And the potential difference on R_L is $V = I\dfrac{R_1 R_L}{R_1 + R_L}$, and substitute I into it to obtain

$$V = \dfrac{\dfrac{R_1 R_L}{R_1 + R_L} V_0}{R_2 + \dfrac{R_1 R_L}{R_1 + R_L}}$$

Numerator and denominator are multiplied by $(R_1 + R_L)$ to obtain

$$V = \dfrac{R_1 R_L V_0}{R_2 (R_1 + R_L) + R_1 R_L} \quad (18-14)$$

For a given power E and load R_L, V is only related to R_1 or R_2, and $R_1 + R_2 = R_0$ is also given (i.e. the total resistance of slide rheostat), so the formula (18-14) can shows the relation between V and slide position. In order to see their relation clearly, suppose $X = R_1/R_0$ and $K = R_L/R_0$. X shows the relative position of slide on resistor, K shows the ratio of load resistor and slide rheostat, substitute it into formula (18-14), and obtain

$$V = \dfrac{XKV_0}{X + K - X^2} \quad (18-15)$$

For different K values, the relation of X and V is shown in Fig. 18-17, the following is to discuss the two special conditions:

①When $K \gg 1$, i.e. $R_L \gg R_0$, note $X \leq 1$, and formula (18-15) can be simplified into

$$V = XV_0 = \dfrac{R_1}{R_0} V_0$$

At the moment V is proportional to X, and the relation is linear.

②When $K \ll 1$, i.e. $R_L \ll R_0$,

$$V = \dfrac{KV_0}{1 - X} = \dfrac{R_L V_0}{R_0 - R_1} = \dfrac{R_L}{R_2} V_0$$

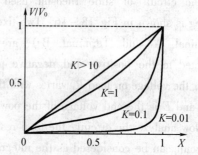

Fig. 18-17 Voltage dividing characteristic curve

V is inversely proportional to R_2. With the decrease of R_2, V starts to increase slowly. When $R_2 \to 0$ (X is close to 1), the change of V quickly increases to the maximum value.

Combining the experiment condition, discuss how to use Fig. 18-17 and formula (18-15) to select rheostat according to the voltage regulating range and requirements on load:

①If V varies uniformly with the change of X within $0 \sim V_0$, the curve of $K > 1$ is more suitable, and the curve of $K < 1$ is not suitable. Therefore, rheostat can be selected according to load value.

②If the end voltage of the power far exceeds the required voltage on load, e. g. the end voltage of power is 2.6 V, adjustable range of rheostat voltage is 0~2.6 V, but the load required voltage is adjusted within 0~0.2 V, at the moment the curve of $K = 0.1$ is more suitable, but be careful when adjusting X, for when making X close to 1, load is easily burned.

③In Fig. 18-17, the part of abrupt change of $K = (0.1 \sim 0.01)$ curve can be used to make sensitive reaction to the critical point in automatic control.

Additionally, the selection of rated current of slide rheostat used as voltage divider should be considered from the maximum value of total current I, and I increases with the monotone increase of R_1. When $R_1 \rightarrow R_0$ ($R_2 \rightarrow 0$), there should be the maximum value of current according to formula (18-13).

$$I_0 = \left(\frac{R_0 + R_L}{R_0 R_L}\right) \cdot V_0 \qquad (18-16)$$

It can be seen from the above formula that: when $R_L \gg R_0$, mainly consider the current passing through R_0; when $R_L \ll R_0$, the current passing through loads R_L and R_2 is very high. If the rated current of rheostat or load is not enough, the instrument is easily burned; therefore, at the moment the selection of the rated current of the slide resistor mainly consider the maximum current passing through R_L and R_2.

(2) The control current characteristic curve of slide rheostat

The circuit of slide rheostat used as control current device is shown in Fig. 18-18. At the moment the current in R_L equals to the output current of the power

$$I = \frac{V_0}{R_L + R_1} \qquad (18-17)$$

Introduce the parameters $X = \frac{R_2}{R_0}$ and $K = \frac{R_L}{R_0}$ to it, so there is

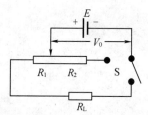

Fig. 18-18 Shunt circuit

$$I = \frac{1}{1 + K - X} \cdot \frac{V_0}{R_0} \qquad (18-18)$$

As shown in Fig. 18-18, suppose $I_0 = \frac{V_0}{R_L}$, and there is

$$\frac{I}{I_0} = \frac{K}{1 + K - X} \qquad (18-19)$$

For different values of K, the relations to X (I/I_0 control current characteristic curve) is as shown in Fig. 18-19.

Discussions on control current characteristics:

①For control current circuit, the current passing through load R_L can not be zero, and the bigger the value of $K = R_L/R_0$ is, the smaller the selected resistance value R_0 of slide rheostat of shunt current is, and the smaller the adjustable range of current is.

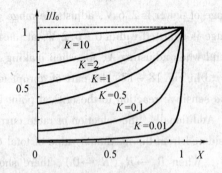

Fig. 18-19 Control current characteristic curve

②For $K\dfrac{R_L}{R_0} \geq 2$, i.e. $R_0 \leq \dfrac{1}{2}R_L$, it can be known that from control current characteristics curve diagram, adjusted linearity is better, i.e. it is easy to be adjusted to the required values of I. For the curve $R_0 \gg R_L$ (such as $K = 0.01$) which is smaller than K, the current change is very small (I/I_0 from $0.01 \rightarrow 0.1$) within a very big section at the beginning (R_2/R_0 from $0 \rightarrow 0.9$); and within a very small range in a rear section, the current change is very big (I/I_0 from $0.1 \rightarrow 1.0$), so it is difficult to be adjusted to the required current value (such as $I/I_0 = 0.5$), which is also called insufficient fine adjustment.

In order to make the current adjustment range bigger and easy to adjust, multiple shunt circuit is generally adopted, and here is no more discussion.

Measuring Resistance by Assembly Bridge

1. Background and application

The bridge circuit is a basic one in electricity. Electrical measuring instruments designed using the bridge balance principle can be used to not only measure resistance, but also capacitance and inductance. By measuring these quantities we can indirectly measure the amount of non-electrical such as temperature, pressure and so on. Therefore, the bridge circuit is widely used in automated instrumentation and automatic control.

The bridge circuits can be divided into DC and AC bridges. DC bridge is an instrument for measuring the resistance and resistance related physical quantity; AC bridge is mainly used to measure physical quantities such as the capacitance and inductance. DC bridges are divided into single bridge and double bridge. DC bridge was invented by Wheatstone in 1843, so called "Wheatstone bridge". Wheatstone bridge is suitable for measuring the resistance range of $1 \sim 10^6$ Ω. DC double bridge (Kelvin Bridge) is suitable for measuring the resistance range of 10^{-5} Ω $\sim$ 10 Ω.

In this study, we will use the Wheatstone bridge to measure the resistance. Wheatstone (1802 – 1875) is a British physicist. (see Fig. 19 – 1) He was born in 1802 in Gloucestershire, England. In his youth, he got strictly formal

Fig. 19 – 1   Charles Wheatstone

training, his interest was extensive, his ability was very strong. In 1834 he was appointed professor of experimental physics by the London King College. In 1836, he was elected as member of the Royal Society of London, and in the next year he was elected as a foreign academician of the French Academy of Sciences. In 1868, he was knighted by the King. He died at the age of 73 in Paris on October 19, 1875. Wheatstone showed great interest in the study of physics very early. He had made important contributions in many areas of physics. In the study of electricity, Wheatstone had many unique methods and insights. He subtly measured velocity of electromagnetic wave in metal conductors using the method of rotating piece, in which the measured value was more than 280,000 kilometers per second. Wheatstone used the relatively large rotating speed quantity instead of the very small time interval, afterwards this method was used by the French physicist Foucault (1819—1868) to accurately measure the speed of light for the first time. Wheatstone was one of the first group of British scientists who truly comprehended Ohm's law and put it into practical applications.

In optics, Wheatstone carried out researches on binocular vision, reflecting stereoscope etc. and elaborated the fundamental causes of visual reliability. He also made correct explanations on physiological optics such as the human eye vision, color vision and so on. Wheatstone also studied the tone transmission on a rigid straight wire and made remarkable achievements. He also confirmed Bernoulli principle of air vibration problems in playing musical instruments by experiments.

2. Experiment purposes

Understand and master the principles and methods of "Wheatstone bridge" to measure resistance by the designed experiments; Learning to use exchange method to eliminate the system error in the bridge; Be able to use the "Wheatstone bridge" to solve simple application problems.

3. Design requirements

(1) State the purpose and significance of the experiment in the preview report; expound the basic principle of the experiment; draw the schematic diagram; write the steps of the experiment; design data recording form.

(2) Connect wires correctly and make sure that the layout of instruments is reasonable; record the type or model and quantity of instruments used; record the experiment phenomenon.

(3) Calculate the resistance of the resistance under test and handle the system error-writing

the calculations; write the expression for the measurement results.

(4) Analyze the experimental results and the experimental phenomena, then draw a conclusion and present some suggestion for improvements.

4. Experiment instruments

The main equipments provided of the experiment are ZX21a DC resistance box (see Fig. 19 - 2), HY1711 - 3S DC power supply (see Fig. 19 - 3), AC5/3DC galvanometer pointer, slide-wire rheostat, 9 hole plug-in square plate, resistance to be measured, switch and meter.

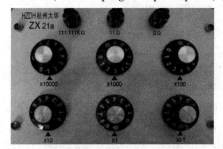

Fig. 19 - 2 ZX21a DC resistance box

Fig. 19 - 3 DC regulated power supply

(1) ZX21a DC resistance box, maximum range (0 ~ 99,999.9 Ω).

(2) HY1711 - 3S DC regulated power supply

The power supply adopts the international advanced suspension stability technology incorporated by stabilizing voltage and steady flow techniques. The output voltage is continuous and adjustable. The output current in steady flow state is continuous and adjustable. Two working conditions—voltage stabilizing and steady flow will transform automatically along with the load. Output has an important function in current limitation, short-circuit protection, and opening automatically after the recovery. The form of output is single or dual channels. Dual power supply can be used individually. It also can be used in series, in parallel and track usage.

(3) AZ19 DC galvanometer

Galvanometer (galvanometer) allows only a weak current, so it should be used with great care to prevent damage. It works like this:

①When using the galvanometer, open the lock button first that the lock button turn from the red rod to the write rod, to see whether the pointer finger to zero. If there is a deviation, gently turn the zero adjustment knob to zero.

②When using the galvanometer, the "power meter" button on the galvanometer is the key switch. It needs to be press intermittently to avoid damaging the galvanometer (see Fig. 19 – 4).

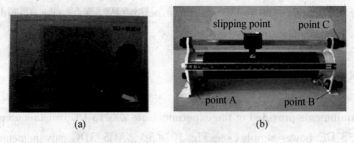

Fig. 19 – 4　AC5/3DC galvanometer
(a) Slide rheostat; (b) Sliding rheostat circuit

(4) Slide-wire rheostat

Slide-wire rheostat is the resistance wire evenly wounding on an insulating porcelain pipe. There are two fixed terminals on the left and right sides of its bottom

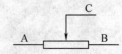

Fig. 19 – 5　Slide-wire rheostat

part. There is a sliding end and the wiper terminal at one end of its upper part (see Fig. 19 – 5). Resistor R is 1,900 Ω, allowing current is 0.3 A. In this study, use slide wire rheostat as the two arms of the bridge.

(5) 9 hole plug-in square plate, resistance, switch and meter

The laboratory provides 9-hole square plug-in plate table, Microampere meter, four resistance: 510 Ω, 1 kΩ, 10 kΩ, 1 MΩ, connecting wire and a power switch. Nine holes in a "field" shape on 9-hole plug-square plate are conducting and disconnected with other holes in different "field" shapes (see Fig. 19 – 6).

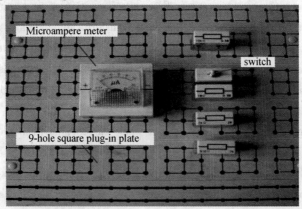

Fig. 19 – 6　9-hole square plug-in plate, resistance, switch and header

5. Experiment principle and method prompt

The working principle of Wheatstone bridge

The basic circuit of single bridge—also known as Wheatstone is shown in Fig. 19 – 7. It consists of four bridge arms, the "Bridge" balance indicator (usually a galvanometer), working power supply E, switch and other components. Select R_1, R_2 suitably, adjust the standard resistor R_s, equalize the potentials of point B and D to make the galvanometer refer to zero, at the moment the bridge is in the status of balance. When the bridge is balanced there is

$$I_1 R_1 = I_2 R_2, I_x R_x = I_s R_s, I_1 = I_x, I_2 = I_s \qquad (19-1)$$

So we can get that $\dfrac{R_1}{R_2} = \dfrac{R_x}{R_s}$, that is

$$R_x = \frac{R_1}{R_2} R_s = k R_s \ (k = R_1/R_2) \qquad (19-2)$$

The formula is called the bridge balance condition. So the essence of using DC bridge to measure the resistance R_x is to compare the resistance R_x with the standard resistance at the known rate k in the bridge balance condition. So the bridge method is also called "balance comparison method".

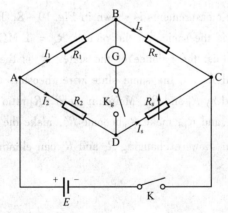

Fig. 19 – 7 Principle of Wheatstone bridge circuit

Sensitivity of Wheatstone bridge

Sensitivity S is defined that the bridge arm resistance R_s change ΔR_s, galvanometer deviate from the equilibrium position of the lattice with Δd in the bridge balance condition. Therefore I_p, easy to prove that U_{G_2K}, so $S = \dfrac{\Delta d}{\Delta R_x / R_x}$. Usually define pointer deflection 0.2 grid as the limitation that eye can perceive, the measurement error introduced by the sensitivity limit is $\Delta R_x = R_x \dfrac{0.2}{S}$.

The sensitivity of the bridge is decided by the size of the supply voltage, internal resistance of galvanometer and bridge arm resistance. When the voltage and the galvanometer sensitivity are high, the sensitivity of the bridge is high. When the resistance galvanometer and the bridge arm resistance are high, the sensitivity of the bridge is low. In theory, the higher the sensitivity of the bridge is, the more accurately the bridge balance can be judged, the smaller uncertainty of measurement results can be controlled. But in fact the higher sensitivity of the bridge is not better in some cases, the higher the sensitivity means that it takes longer time to adjust the balance, thus the stability and repeatability is poor and the experiment operation is inconvenient. Therefore you should choose reasonable power supply voltage, galvanometer and corresponding bridge arm resistance according to actual situation, moderately improve the sensitivity of the bridge and at the same times the experimental requirements.

Exchange measurements (reciprocity law)

The circuit of exchange measurements is shown in Fig. 19 – 8. Fig. 19 – 7 is a modification of this, in which the effect is the same. In the picture $R_m = 1$ MΩ, its role is to protect the galvanometer and easy to adjust the balance of the state. R_s is the resistance box, R_x is the measured resistance, R_1 and R_2 are the same slide wire rheostat. Exchanging R_x and R_s can eliminate the error introduced by R_1 and R_2. Maintain the R_1/R_2 ratio under the same conditions, exchange the position of R_s, and R_x, make R_s become R_s', make the bridge balance again, then $R_x = \sqrt{R_s \times R_s'}$. The equation shows exchanging R_x and R_s can eliminate the error introduced by R_1 and R_2.

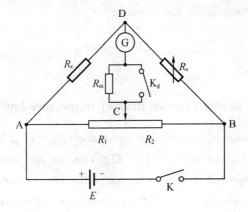

Fig. 19-8 Wheatstone bridge circuit switching method

6. Experiment content

(1) Assembly Wheatstone bridge with the instruments provided by laboratory. Measure the nominal value of 510 Ω, 1 kΩ, 10 kΩ resistance with exchange measurements. Each resistor should be measured five times. Design the record form by yourself. Calculate the uncertainty of the resistance under test according to the accuracy level of the resistance box and write the expression of the resistance measurements.

(2) Use Wheatstone bridge to measure the head resistance of microampere meter without galvanometer. Prompt: use header as measured arm and bridge balance indicator.

7. Analysis and questions

(1) Why use a bridge to measure resistor instead Voltammetry or ohmmeter?

(2) Try to prove: When using exchange method to measure resistance, $R_x = \sqrt{R_s \times R_s'}$, R_s is the value of the comparison arm in the bridge balance at the first time. R_s' is the value of the comparison arm in the bridge balance at the second time after exchanging R_x and R_s.

(3) Do the following factors increase the measurement error of Wheatstone bridge?

①Power supply voltage is instable.

②Resistance of the wire on the ratio arm can not be ignored.

③Galvanometer is not referred to zero.

④Galvanometer sensitivity is not high enough.

8. Appendix

Principle of using Kelvin Bridge (Double Bridge) to measure low resistance

When using single bridge to measure low resistance values of a few ohms, lead resistance and contact resistance R (about $10^{-2} \sim 10^{-4}\,\Omega$) can not be ignored, resulting in greater measurement error. The improved method is to change the low resistance bridge arm to the four terminal method, and then add with a pair of high resistance. After using four-wire connection, the equivalent circuit is Fig. 19-9. R_1, R_2, R_3, R_4 are the wires and the contact resistances. R_1, R_4 are in series with the power supply circuit, the effect is negligible. R_2, R_3 are connected to the high resistance, the effect is negligible.

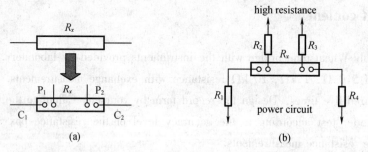

Fig. 19-9 Principle of four-wire connection

(a) Four-wire connection; (b) Equal circuit of four-wire connection

Based on this principle, double bridge is shown in Fig. 19-10. Obtained by the circuit equations, $R_x = \dfrac{R_2}{R_1}R + \dfrac{rR_1'}{R_1' + R_2' + r}\left(\dfrac{R_2}{R_1} - \dfrac{R_2'}{R_1'}\right)$. Make R as small as possible and make the two contrast ratio arm into the linkage mechanism. Try to make $\dfrac{R_2'}{R_1'} = \dfrac{R_2}{R_1}$, so $R_x = \dfrac{R_2}{R_1}R = kR$.

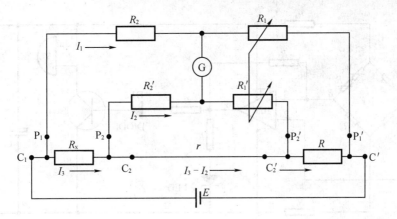

Fig. 19 – 10 The principle of the double bridge

Temperature alarm circuit design based on Wheatstone bridge

With the development of computer and sensor technology, bridge circuit has many applications in manufacturing and research in particular aspects of measurement and automation. The design is based on a Wheatstone bridge to achieve temperature alarm function. Temperature alarm circuit mainly is composed by the power supply, a Wheatstone bridge (including temperature sensors), comparator, alarm device components. Fig. 19 – 11 is a circuit diagram. The temperature sensor R_t is a negative temperature coefficient thermistor, W_t is a potentiometer. R_1, R_2, R_3 are equal-value resistors. R_1, R_2, R_3, $(R_t + W_t)$ consist of a Wheatstone bridge; comparator is constituted by the integrated chip IC1; Warning means is consisting of a light emitting diode D_0, buzzer composition Y. Adjust W_t to make R_t and W_t value is greater than the sum of R_3, thus the potential at point B is less than A. The terminal phase of potential IC1 is lower than the inverting input, output low level D1, BG1 cut off. Put R_t in the environment in which the temperature is rising (such as heating water). When the temperature rises, R_t value decreased, the pressure drop of R_t also reduces, thus the potential at point B rises. When the potential at point B is higher than that at point A, IC1 output a high level, D1, BG1 conduct, light-emitting diode D_0 start, Y speaker sound an alarm, it show that the temperature exceeds the set value. Adjustment potentiometer W_t can set different alarm temperature.

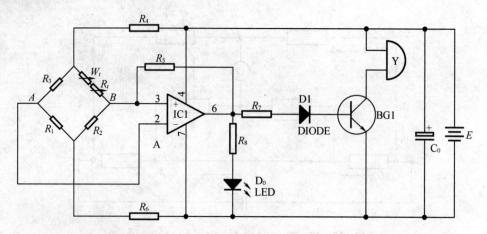

Fig. 19−11 The principle of the double bridge temperature alarm circuit principle diagram

In the experiment, select MF51 negative temperature coefficient thermistor as R_t, may choose LM393 as IC1, and choose the regulated power supply provided by the experiment as the power. Heating thermostat street lamps light control and other function can be achieved on the basis of this circuit.

Experiment 20

Assembly Telescope and Microscope

1. Background and application

Telescope

In the early 17th century, Hans Lippershey, the owner of an optical shop in the Dutch town, arranged in order of a convex lens and a concave lens to check the quality of the polished lens. He found that the distant church spire was seemingly brought nearer when viewed through the pair of lenses. So he discovered the telescope accidentally. In 1608, he applied for a patent for his telescope, and made a binocular microscope. It is said that dozens of glasses artisans had claimed the invention of the telescope, but Lippershey is generally considered as the inventor of the telescope.

News of the invention of the telescope quickly spread through Europe. Italian scientist Galileo (Galileo Galilei) also made a telescope after hearing the news. The first telescope was only able to magnify objects three times. A month later, he produced a second telescope could magnify 8 times, a third telescope could magnify up to 20 times. October 1609, he made a telescope could magnify 30 times (see Fig. 20 – 1). Galileo used a self-made telescope to observe the night sky, first discovered the surface of the moon to be not smooth and is

Fig. 20 – 1 Galileo' telescope

covered with mountains and shield volcanoes. Later it was found that Jupiter has 4 satellites in orbit around it, the sunspot activity, thereby confirming that the Sun rotates. Almost at the same time, the German astronomer Kepler (Johannes Kepler) also began to study the telescope. He proposed another astronomical telescope that was composed of two convex lenses, which had larger view field than that of Galileo telescope, but Kepler did not assemble this telescope. During 1613—1617, Scheiner firstly constructed this telescope. According to Kepler's advice, he made the telescope using three convex lenses to show the upright image instead of the inverted image formed by two convex lenses. Christoph Scheiner constructed eight telescopes and observed the sun using his telescope one by one. No matter which telescope he could see sunspots of the same shape. Thus he made a lot of people gave up the thought that sunspots may be an illusion caused by dusts on the lens. He proved the existence of sunspots is indeed true. When observing the sun, Scheiner used special shading glass, but Galileo did not, as a result his eyes were hurt and finally almost blind. To reduce the chromatic aberration of refracting telescopes, Netherlands Huygens (Christian Huygens) made a tube length of nearly 6-meter telescope to observe Saturn's rings in 1665. Later he produced a nearly 41m long telescope.

The telescope using lens as an objective lens is called refracting telescope. Even longer tube, precision machining of the lens, chromatic aberration can not be eliminated. Newton also considered refractor chromatic aberration was incorrigible, but in 1668 he invented reflecting telescope to solve the problem of chromatic aberration. The first reflecting telescope was very small, the interior diameter was only 2.5 cm, but was able to clearly see the moons of Jupiter, Venus phases and so on. In 1672, Newton produced a larger reflecting telescope and gave it to the Royal Society. It is still preserved in the library of the Royal Society (see Fig. 20 - 2). In 1733, British Hall (Chester Moore Hall) made the first achromatic refractor. In 1758, John Dollond in London also made the same telescope. He adopted different refractive index of glass to manufacture convex and concave lenses, respectively, in which the formed color edges were counteracted with each other. However, a large lens is not easy to manufacture. Currently the diameter of the world's largest refracting telescope is 102 cm, installed in the Yerkes observatory. In 1793, British Herschel (William Herschel) produced a reflective telescope, in which the reflective mirror with a diameter of 130 cm was made of copper-tin alloy and weighed one ton. In 1845, British Parsons (William Parsons) made a reflecting telescope with the reflecting mirror with a diameter of 1.82 m. In 1917, Hooker Telescope (Hooker

Fig. 20 - 2 Newton's telescope

Telescope) was built in California's Mount Wilson Observatory. Its main reflector diameter was 100 inches. It is using this telescope that Hubble (Edwin Hubble) discovered the amazing fact that the universe is expanding. In 1930, the German Schmidt (Bernhard Schmidt) combined the advantages of refracting telescope and reflecting telescope to produce the first catadioptric telescope (Refracting telescope aberrations are small but has chromatic aberration. The bigger the size is, the more expensive the refracting telescope is; the reflecting telescope does not have chromatic aberration, it can be made bigger and the cost is low, but there are aberrations.)

After the Second World War, reflecting telescopes developed quickly in the astronomical observation fields. In 1950, Hale reflecting telescope was installed in Palomar Observatory. A reflecting telescope with 6 m diameter was installed in 1969 in the northern Caucasus Mountains Pasto Hove, the former Soviet Union. In 1990, NASA sent Hubble Space Telescope into its orbit. However, for there was something wrong with the mirror, it was not until 1993 that the Hubble telescope began to fully work after astronauts completed the repair and replacement of the lenses in space. Because it can not be interfered by the earth's atmosphere, the images delivered by Hubble telescope are 10 times clearer that those by the same telescope on earth. In 1993 the United States built "Keck telescope" in which the diameter is 10 m on Mauna Kea in Hawaii. Its mirror plane is comprised of 361. 8 m reflecting mirrors. In 2001, the European Southern Observatory in Chile completed the construction of "Very Large Telescope" (VLT). It consists of four 8 m-diameter telescope, whose light-gathering power is equal to a 16 m reflector telescope.

Astronomical telescope is an important means of observing astronomical objects. Without telescope's advent and development, there is no modern astronomy. With the improvements of the telescope in various aspects, astronomy is also experiencing a striding development and rapidly promoting human understanding of the universe.

Further, another important use of telescopes is for military. Military telescope is mainly used for the observation of battlefield, terrain and reconnaissance etc.

Microscope

As early as the 1st century BC, it was found that the amplified image was given when observing tiny objects using the spherical transparent objects. Then people gradually understood the fact that spherical glass surface can enlarge the image of objects. In 1590, the Netherlandish and Italian glass makers created enlargement instruments similar to the microscope. Around 1610, when Italian Galileo and German Kepler were studying telescopes, they changed the distance between the objective lens and the eyepiece and proposed the reasonable microscope optical structure. Optical manufacturer engaged in the manufacture, promotion and improvement of microscopes.

In mid-17th century, British Hooke and Dutch Leeuwenhoek both made outstanding contributions to the development of the microscope. Around 1665, Hooke added coarse and focusing mechanism, lighting systems and bearing specimens piece bench into microscope. These components become a fundamental part of modern microscopes through continuous improvement. In the period 1673—1677, Leeuwenhoek created at least 25 microscopes, of differing types, of which only nine survived. Hooke and Leeuwenhoek made outstanding achievements in the study on microstructures of animals and plants.

In 19th century, the emergence of high-quality achromatic immersion objective lens greatly improved the microscope ability to observe microstructures. In 1827, Amici first used the immersion objective. In 1870s, the Germans Abbe created the classical theoretical foundation of microscopy imaging. These had promoted the rapid development of manufacturing technology and microscopic observation of microscopes, which provided a powerful tool for biologists and medical scientists including Koch, Pasteur etc. to discover bacteria and microbes in the second half of the 19th century.

While the structure of microscopes is developing, microscopic observation techniques are constantly innovating. In 1850, polarizing microscopy was created; in 1893, interference microscopy was created; in 1935, Dutch physicist Zernike (Frits Zernike) created phase contrast microscopy and he won the Nobel Prize in Physics in 1953.

Classical optical microscope was just a combination of optical component and precision mechanical components; it used the human eye as a receiver to observe the magnified image. Later in the microscope added with cameras, the film can be as a receiver to record and store images. Modern optoelectronic devices, television camera tubes and charge-coupled devices are widely used as a receiver, with the help of the computer to construct image information acquisition and processing system.

Ruska (Ernst Ruska, Germany), Binney (Gerd Binney, Germany) and Rohrer (Heinrich Rohrer, Switzerland) shared the Nobel Prize in Physics in 1986 for invention of the electron microscope and scanning tunneling microscopy. Hundreds of years ago, Leeuwenhoek regarded his microscope technology as a secret. Today, the microscope (at least an optical microscope) has become a very common tool to let us understand this microscopic world (see Fig. 20-3).

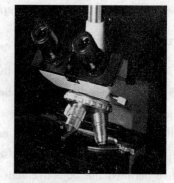

Fig. 20-3 Modern optical microscope

Assembly Telescope and Microscope Experiment 20

In recent years, with the rapid development of science and technology, microscopy is applied more widely and more types of microscopes appear. It can be used for the transmitted light, reflected light, and bright field microscopy, dark field, phase contrast, differential interference contrast, fluorescence, polarization, etc, based on the use of optical principles. It also can be equipped with various accessories, such as microscopic photography, TV, projectors, and temperature control object stage. It is able to be widely used in anatomy, biology, bacteriology, histology, pharmacology, biochemistry, geology, micro-fiber science, soil research, industrial production, leather industry, metallography, neurology, osteopathy, physiology, radiology, serology, veterinary sciences, and pollution studies and so on.

2. Experiment purposes

To deeply understand the rule of thin lens imaging through the experiments; to understand the basic structure and working principle of the telescope and microscope; to learn the analysis and adjustment method of simple optical path; to learn the method to measure the focal lengths of lenses.

3. Design requirements

(1) Assemble a simple structured telescope and microscope according to the basic structure and principles of the telescope and microscope, using equipments provided by the laboratory.

(2) Design a reasonable optical path diagram to measure the focal length of a convex lens and a concave lens.

(3) Design an experimental data recording form.

4. Experiment instruments

The main experimental equipments available are given below: convex lens, concave lens, object screen, image screen (reticle), optical bench with the scale, etc. (see Fig. 20-4).

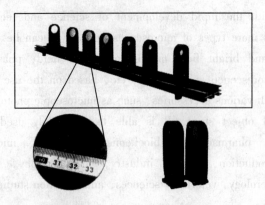

Fig. 20-4　Experimental device

5. Experiment principle and method prompt

Magnifying lens and Magnification

Eyepiece is used in both microscopes and telescopes. Eyepiece is actually a simple magnifying lens. It can magnify an object to be observed, the most important parameter is the angular magnification m.

Here is a concept of magnification. Magnification is magnification times. It is the ratio of the size of original object to the size of final image seen by the human eyes after being amplified by the objective lens and the eyepiece. It is the product of the magnification of the objective lens and the eyepiece. There are two kinds of magnification: When the lens or spherical mirror image, the ratio of the image height to the original height is called the line magnification (for cameras, projectors and other optical instruments); When using visual instruments to observe objects, the ratio of the image-to-eye opening angle (view angle) to the object-to-eye one of the direct observation is called the angular magnification. Angular magnification of the telescope is equal to the ratio of the focal length of the objective lens to the eyepiece. Angular magnification of convex lens is approximately equal to the ratio of the near point distance (about 25 cm, an estimate of the "near point" distance of the eye—the closest distance at which the healthy naked eye can focus) to the focal length when used as a magnifying lens. For example, the focal length of a magnifying lens is 10 cm and its angular magnification is 2.5 times, usually written 2.5x. The angle magnification of the microscope is equal to the product of the line magnification of objective lens and the angle magnification of eyepiece. The objective lens and the eyepiece of microscope are engraved 40x (the line magnification of objective lens) and 10x (the angle magnification of eyepiece). You can make microscopes with varying angular magnification with combinations of different objective lens and eyepieces.

Microscope

Microscope can observe small objects and it is optical instrument for measuring small distances, the optical path as shown in Fig. 20 - 5. Focal length of the objective lens L_0 (typically less than 1 cm) is very short, the eyepiece focal length L_e is longer than the focal length of the objective lens, but not more than a few centimeters. The distance between reticle P and the objective lens L_0 is l. Put the object screen y on an arbitrary point outside of the focus F_0 (see Fig. 20 - 5), adjust the distance between y and L_0 to make a zoom, inverted real image y' in the reticle P through lens L_0. Then observe the image y' through the eyepiece L_e, first adjust the distance between eyepiece L_e and reticle P to watch clear reticle P, then see y' clearly, and also see the reticle P clearly. And eyepiece L_e plays a role of magnifying, it turns y' as a magnified virtual image y'' (Reticle P is simultaneously amplified to be a virtual image P', and coincides with y''). Then the tiny objects observed by human eyes are greatly amplified into y''.

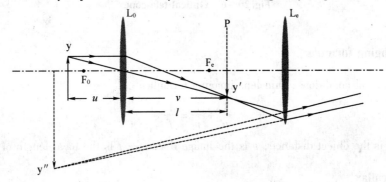

Fig. 20 - 5 Optical microscope

Telescope

A telescope is divided into Kepler and Galileo telescope. Objective lens and eyepiece of the Kepler telescope are convex lenses, while the objective lens of Galileo telescope is convex lens, eyepiece is concave lens. Take the Kepler telescope for example, we describe the basic working principle of the telescope.

The optical path of the Kepler telescope is shown in Fig. 20-6, The light (parallel light) of a point on the object screen in infinity (not shown) via the objective lens L_0 make a real image y' on the focal plane of L_0 (inside the focus F_e of the eyepiece L_0), the reticle P is also located at the focal plane of L_0, and then y' coincides with the reticle P. If y is not in the infinity, then y' and P is outside of the F_0. The observation process is the same as the microscope when observing y'' with human eye through the eyepiece L_e. When people observe objects through a telescope, it is equivalent to make the distant objects near to observe, it actually plays a role in magnifying view angle.

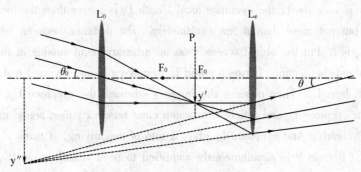

Fig. 20-6 Optical telescope

Thin lens imaging formula

Under paraxial conditions, thin lens imaging formula

$$\frac{1}{u} + \frac{1}{v} = \frac{1}{f} \qquad (20-1)$$

Where u is the object distance; v is the image distance; f is the focal length of the lens.

Eliminate parallax

In optical experiments, we often accurately measure the image size, location, etc. In the adjustment process we must pay attention to eliminate parallax. The reason for parallax is: If the reticle and the image are not coplanar, with shaking our eyes (viewing position slightly changes), relative movement will take place between the reticle and the image. If there is a parallax between the reticle and the image, meaning that they are non-coplanar, we should slightly adjust the position of the reticle or the image along with shaking our eyes slightly until there is no relative movement between the reticle and the image. In this case, the image locates on the reticle and its position accurately determines.

6. Experiment content

(1) Assembly an infinitely focused telescope by yourself

Take use of a long focal length convex lens, a short focal length convex lens, a reticle with crosshairs, and an object screen provided by the laboratory, to assemble an infinitely focused telescope. Please draw the experimental optical path diagram, record experimental data.

Tip: The characteristic of the infinitely focused telescope is that the distance between the reticle and the objective lens is equal to the focal length of the objective lens, so try to accurately measure the focal length of the objective lens.

(2) Measure another focal length of convex lens with the assembly infinitely focused telescope by yourself.

Please draw the experimental optical path diagram, recording experimental data.

Tip: The telescope is infinitely focused, when observing objects with it, the incident light is required to be parallel light, otherwise the object is not clear, so the light emitted by a point of object screen into the telescope through measured lens must be parallel.

(3) Measure the focal length of concave lens with the assembled infinitely focused telescope by yourself.

Please draw the experimental optical path diagram, recording experimental data.

Tip: You can perform an experimental operation on the basis of the experimental content mentioned above.

(4) Assembly microscope by yourself.

According to the principle of the microscope, select convex lens with the shortest focal length as the objective lens and the other convex lens with short focal length as the eyepiece in the optical components. In the experiment, we can change the magnification of the microscope by changing the position of the objective lens. The experimental content is a self-assembly and observational experiment, it does not require a quantitative measurement.

(5) Experiment summary, writing experimental report.

7. Analysis and questions

(1) Where is the major difference between Kepler telescope and microscope in the basic

structure?

(2) Galilean telescope consists of a convex lens (objective) and a concave lens (eyepiece), we can see an upright virtual image of the object. Please answer how to install the lens and draw optical path diagram according to the law of the lens imaging.

参 考 文 献

[1] Richard P F, Robert B L, Matthew Sands. The Feynman Lectures on Physics, boxed set: The New Millennium Edition [M]. New York: Basic Books, 2011.

[2] 孙晶华,梁艺军,关春颖,等. 操纵物理仪器,获取实验方法——物理实验教程[M]. 北京: 国防工业出版社, 2009.

[3] 邱春蓉, 黄整. 大学物理实验双语教程[M]. 成都: 西南交通大学出版社, 2010.

[4] 霍剑清. 大学物理实验（第一册 — 第四册）[M]. 北京: 高等教育出版社, 2002.

[5] 唐晋生, 吴宗森, 盛克敏. 大学物理实验(双语教学用书)[M]. 北京: 国防工业出版社, 2011.

[6] 沈元华, 陆申龙. 基础物理实验[M]. 北京: 高等教育出版社, 2003.

[7] 朱鹤年. 基础物理实验教程: 物理测量的数据处理与实验设计[M]. 北京: 高等教育出版社, 2003.

[8] 周殿清. 大学物理实验教程[M]. 武汉: 武汉大学出版社, 2005.

[9] 杨燕. 华侨华人留学生高等教育系列精品教材·大学物理实验[M]. 2 版. 广州: 暨南大学出版社, 2010.

[10] 张晓波, 李小云. 大学物理实验[M]. 杭州: 浙江大学出版社, 2008.

[11] 苑立波, 梁艺军, 杨军, 等. 光纤实验技术[M]. 哈尔滨: 哈尔滨工程大学出版社, 2005.

[12] 孙晶华, 梁艺军. 大学物理实验[M]. 哈尔滨: 哈尔滨工程大学出版社, 2008.

[13] 王雪银, 陶纯匡, 汪涛, 等. 大学物理实验[M]. 北京: 机械工业出版社, 2005.

[14] 陆廷济, 胡德敬, 陈铭南. 物理实验教程[M]. 上海: 同济大学出版社, 2003.

[15] 黄志高. 大学物理实验[M]. 北京: 高等教育出版社, 2007.

[16] 刘战存, 郑余梅. 霍尔效应的发现[J]. 大学物理, 2007, 26(11): 51 - 55.

[15] 钱锋、潘人培. 大学物理实验[M]. 北京: 高等教育出版社, 2005.

[17] 丁慎训, 张连芳. 物理实验教程[M]. 2 版. 北京: 清华大学出版社, 2002.

[18] 杨述武, 赵立竹, 沈国土, 等. 普通物理实验[M]. 北京: 高等教育出版社, 2007.

[19] 高立模. 近代物理实验[M]. 天津: 南开大学出版社, 2006.

[20] 耿完桢, 金恩培, 赵海发, 等. 大学物理实验[M]. 哈尔滨: 哈尔滨工业大学出版社, 2005.

[21]赵梓森.光纤通信工程[M].北京:人民邮电出版社,1998.

[22]李金海.误差理论与测量不确定度评定[M].北京:中国计量出版社,2003.

[23]郭奕玲.物理学史[M].北京:清华大学出版社,2005.

[24]赵文.采用模拟基带调制的光纤视频传输[J].光通信技术,1980,3(3):18-27.

[25]罗志高,郑兴世.模拟脉冲宽度调制(PWM)信号光纤传输实验系统[J].实验技术与管理,1998,15(1):28-30.

[26]王玉清.几种测量惯性质量的方法[J].大学物理实验,2008,1:34-37.

[27]徐英勋.毛细管法测量水银的表面张力系数[J].物理实验,2001,1:41-42.

[28]衡耀付,张宏.转筒法测定液体粘滞系数实验的改进[J].天中学刊,2002,5:65-66.

[29]刘文鹏,张庆礼,殷绍唐.粘度测量方法进展[J].人工晶体学报,2007,4:381-404.

[30]鲍修增,王岚.利用毛细管测量血液粘度的研究[J].中国医学物理学杂志,2004,2:102-103.

[31]张彩霞.对空气比热容比测定实验的研究[J].太原师范学院学报,2005,4(1):56-59.

[32]张海涛.霍尔效应及应用[J].温州职业技术学院学报,2005,5(4):26-28.

[33]刘战存,郑余梅.霍尔效应的发现[J].大学物理,2007,26(11):51-55.